THE AI-POWERED SALES REVOLUTION

How Top Performers Are Using AI
to Supercharge the Sales Process

Nicholas McMenemy

SANDHOLME PUBLISHING

SYDNEY, AUSTRALIA

Nicholas McMenemy / Sandholme Publishing

396 George Street, Sydney, NSW, 2026

www.sandholme-publishing.com

Ordering Information:

Quantity Sales. Special discounts are available on quantity purchases by corporations, associations, and others. For details, contact the "Special Sales Department" at the address above.

The AI-Powered Sales Revolution / Nicholas McMenemy—1st ed.

ISBN 978-1-7645551-0-4 (Paperback)

ISBN 978-1-7645551-1-1 (E-Book)

Until one is committed, there is hesitancy,
the chance to draw back, always ineffectiveness.
Concerning all acts of initiative (and creation),
there is one elementary truth,
the ignorance of which kills
countless ideas and splendid plans:
that the moment one definitely commits oneself,
then Providence moves too.

All sorts of things occur to help one
that would never otherwise have occurred.
A whole stream of events issues from the decision,
raising in one's favour all manner of unforeseen incidents
and meetings and material assistance,
which no man could have dreamt
would have come his way.

Whatever you can do, or dream you can,
begin it.

Boldness has genius, power, and magic in it.

— W.H. Murray
from The Scottish Himalayan Expedition (1951)
with a couplet often attributed to Johann Wolfgang von Goethe

Contents

Preface

This book exists because of a simple observation: most salespeople are working too hard for too little return.

I've spent over two decades in technology sales, working across telecommunications giants and venture-backed startups, leading teams from Sydney to San Francisco. I've closed seven-figure enterprise deals and built sales teams from scratch. And throughout that journey, I've watched countless talented salespeople burn themselves out chasing prospects who were never going to buy, crafting proposals that missed the mark, and losing deals they should have won. Not because they weren't good enough, but because they were drowning in the wrong work.

Here, the arrival of Artificial Intelligence (AI) hasn't just changed the game, *it's rewritten the rules entirely*. But here's what most people miss: AI isn't a *replacement* for sales professionals. <u>It's a force multiplier for those smart enough to use it properly.</u>

Consider that this book is not about robots taking over sales. It's about you reclaiming your time, your energy, and your competitive edge. It's about working smarter while your competitors work harder. It's about using technology to do what you've always known mattered most. Building genuine relationships, solving real problems, and creating value that customers care about.

I wrote this for the sales professional who's tired of generic advice and fluffy theory. You won't find motivational platitudes here. What you will find is a practical, step-by-step framework for using AI to transform every stage of your sales process. From identifying the right prospects to closing deals faster and building long-term customer relationships.

Why This Book, Why Now

We're at an important moment in the sales business. Just five years ago, the tools we now possess would have seemed like straight out of a movie. AI can now analyse customer sentiment in real-time, predict which deals are likely to close, generate personalised content at scale, and surface insights from millions of data points in seconds. The question is no longer whether AI will reshape sales. It's whether you'll be among those who tap into it or those who get left behind.

But here's the problem: most books and courses on AI in sales fall into one of two camps. They're written either by technologists who've never actually carried a quota, filling pages with theoretical frameworks that collapse on contact with real customers. Or they're superficial overviews that tell you AI is important without showing you *how* to use it to hit your numbers.

I wanted to write something different. Something that bridges the gap between strategy and execution. Something that respects your intelligence and your time. You don't need another book telling you that personalisation matters or that you should focus on high-value activities. You need a blueprint for doing it. Consistently, systematically, and at scale.

The QUANTUM Framework

At the heart of this book is the QUANTUM Framework. A methodology I've developed and refined over years of implementing a range of tools in high-stakes sales environments.

The QUANTUM Framework consists of seven integrated stages: Qualify (finding prospects worth pursuing), Understand (developing deep customer intelligence), Anticipate (predicting objections and risks),

Work through (orchestrating complex deals through organisational politics), Tailor (personalising at scale), Unify (integrating systems and data), and Measure (tracking what predicts success). Each stage builds on the others, creating a continuous cycle of improvement

In Chapter 2, you'll see how the Qualify stage uses AI to help you escape what I call the "activity trap". That dangerous illusion that more emails, more calls, and more meetings equal more revenue. I'll show you how to use predictive analytics and pattern recognition to identify prospects who match your ideal customer profile, so you can stop wasting time on deals that were never going to close.

The Understand stage, covered in Chapter 3, tackles one of sales' oldest challenges: truly grasping what drives your customer's decisions. You'll learn how AI-powered research tools can surface insights about organisational dynamics, competitive pressures, and individual motivations that would take traditional methods weeks to uncover. More above all, you'll learn how to synthesise that information into actionable intelligence rather than just accumulating data.

Chapter 4 explores the Anticipate phase. Using AI to predict objections, identify risks, and prepare responses before you ever walk into the meeting. This is where amateur sellers get blindsided and professionals demonstrate mastery. I'll walk you through specific techniques for scenario planning and objection mapping that have saved deals at the eleventh hour.

Navigation, which we dive into in Chapter 5, is about steering complex sales cycles through organisational politics and decision-making mazes. AI can help you map stakeholder networks, understand influence patterns, and identify the hidden decision-makers who often kill deals from the shadows. You'll see real examples of how this works in enterprise environments where the org chart bears little resemblance to actual power structures.

Critically, the Tailor stage in Chapter 6 addresses the personalisation paradox: how do you customise your approach for each prospect without spending all your time on content creation? I'll show you how to use AI to generate genuinely personalised proposals, presentations, and communications that land with specific audiences. While maintaining quality control and brand consistency.

Unify, covered in Chapter 7, tackles one of the biggest headaches in modern sales: the explosion of tools, data sources, and communication channels. You'll learn how to create an integrated AI-powered sales ecosystem where insights flow naturally between systems, eliminating redundant work and ensuring nothing falls through the cracks.

Finally, the Measure component in Chapter 8 goes beyond vanity metrics to help you track what predicts success. You'll discover how to use AI analytics to identify leading indicators, optimise your sales motions, and make data-driven decisions about where to invest your time and energy.

What Makes This Different

The QUANTUM Framework isn't academic theory. It's battle-tested methodology that my teams and I have used to consistently outperform targets in highly competitive markets. I've made every mistake described in these pages, learned from each one, and distilled those lessons into systems that work in the real world.

Throughout the book, you'll find specific tools, prompts, and techniques you can implement immediately. In Chapter 9, I provide detailed guidance on building your AI toolkit, including how to evaluate tools, avoid common pitfalls, and create workflows that compound your effectiveness over time. Chapter 10 tackles the advanced strategies—multi-threaded selling, account-based approaches, and complex deal orchestration—that separate top performers from the rest of the pack.

But perhaps most critically, Chapter 11 addresses what everyone's thinking but few are discussing openly: the ethical dimensions of AI-powered sales. How do you maintain authenticity when using automation? Where's the line between personalisation and manipulation? How do you build trust while applying technology? These aren't just philosophical questions. They're practical challenges that will define your reputation and long-term success.

Who This Book Is For

Whether you're a solo consultant looking to punch above your weight, a sales manager trying to scale your team's performance, or an experienced enterprise seller who knows there must be a better way. This book is for you. I'm not interested in impressing you with jargon or complexity. I'm interested in giving you tools you can implement on Monday morning.

If you're in B2B sales, particularly in complex, high-value environments, you'll find the frameworks directly applicable. If you're in transactional sales, you'll discover ways to create use and move upmarket. If you're leading a sales team, you'll gain insights into how to coach your people through this technological transition without losing the human elements that make great sellers great.

The salespeople who thrive in the next decade won't be those who resist AI or those who blindly adopt every new tool. They'll be the ones who understand how to combine human judgment with machine intelligence, who know when to automate and when to personalise, who use technology to amplify their strengths rather than paper over their weaknesses.

How to Use This Book

You can read this straight through or jump to the chapters that address your most pressing challenges. Each chapter builds on previous concepts, but I've written them to stand alone if you need to solve a specific problem quickly. The case studies throughout are real situations I've encountered. Some from my own experience, others from sales professionals I've worked with and coached. Names and details have been changed, but the lessons are authentic.

At the end of most chapters, you'll find implementation guides and specific exercises. Don't skip these. The difference between reading about AI-powered sales and transforming your results is execution. I've designed these exercises to be practical and immediately valuable, not busy work.

You're holding a playbook for becoming the kind of sales professional who doesn't just survive the AI revolution but dominates because of it. The future of sales isn't about choosing between human touch and technological power. It's about integrating both so smoothly that your competitors won't know what hit them.

Let's get to work.

CHAPTER 1

THE GREAT SALES DISRUPTION

By the end of this chapter, you will be able to:

- Explain why traditional selling is failing as buyers complete evaluation work earlier. Often using structured frameworks and AI-assisted scoring before a salesperson is meaningfully involved.

- Diagnose what's really changed in deal outcomes: the loss of information asymmetry, process use, and narrative authority. And the resulting commercial impact (longer cycles, higher discounting, more late-stage losses).

- Shift your approach from persuasion to evidence by understanding the four filters buyers now use to reduce uncertainty: risk predictability, economic certainty, peer validation, and time-to-value.

- Apply the Sales Relevance Stack (Signal → Insight → Risk Framing → Economic Clarity) to spot where your deals are becoming interchangeable. And what to change first.

- Pressure-test your current sales motion using the Sales Relevance Test—so you can defend price with evidence and intervene earlier before deals stall, shrink, or get de-scoped.

Why Traditional Methods Are Failing

Nick ran a textbook enterprise sales cycle and still lost the deal - dozens of calls, multiple site visits, a committed internal champion, and a clear path on a $850,000 software purchase that, on paper, should have been straightforward for an experienced seller.

On the ground, the loss didn't come with drama. It came as a decision already finalised. Shaped less by the executives Marcus had spent months influencing and more by a junior analyst running an AI-assisted evaluation. The scoring model carried more weight than the relationships.

That's the reality of modern sales: effort still matters, but it's no longer the differentiator. Decision advantage is, and increasingly, decision advantage comes from how information is gathered, analysed, and compared. Often this is done by systems, or technology and no longer solely by people.

The Structural Shift Most Sellers Miss

The challenge in modern selling isn't effort. It's timing and draw on. It has become harder because the mechanics of buying have changed faster than most sales teams have adapted. In many organisations, the sales teams are running last decade's playbook against a different buying system.

For decades, enterprise selling benefited from three structural advantages:

- Information asymmetry: sellers typically knew more than buyers.
- Process apply: sellers could shape the sequence and pace of the buying journey.
- Narrative authority: sellers largely controlled how value, risk, and trade-offs were framed.

Those advantages have weakened or disappeared. Today, buyers often engage after substantial work is already done:

- Research has been completed through independent sources and peer networks.

- Internal alignment is already forming across finance, security, procurement, and the business.
- Benchmark pricing and viable alternatives are already established.
- Risk has been assessed—often with structured frameworks, data, and increasingly, AI-assisted evaluation.
- The commercial impact is measurable: cycle times extend, discount pressure rises, and late-stage losses increase because sellers are brought in after value bands and shortlists are effectively set.

Gartner's research consistently points to the same pattern: buyers complete a significant portion of their decision-making before engaging a salesperson. [1]

From Persuasion to Signal

Sales authority has shifted away from persuasion and toward evidence. The signals buyers use to reduce uncertainty.

In practice, buyers increasingly assess vendors through a consistent set of filters:

- Risk predictability: the likelihood of implementation failure or operational disruption.
- Economic certainty: total cost versus the outcomes the business can reasonably expect.
- Peer validation: proof that comparable organisations have adopted the solution successfully.
- Time-to-value: the speed at which benefits are realised, not merely promised.

AI is now embedded in that filtering process. It is used to standardise comparisons, identify risk patterns, and pressure-test assumptions.

McKinsey[1]'s research indicates that organisations applying AI to commercial decision-making can improve speed and decision quality. Often alongside higher confidence in the outcome. [2]

When sellers can't add insight beyond these signals, they are treated as interchangeable. And priced accordingly.

Mini-Framework: The Sales Relevance Stack

In modern enterprise buying, sales conversations tend to succeed or fail across four stacked layers. When one layer is missing, commercial outcomes deteriorate. Cycle times extend, discounting increases, and differentiation collapses.

- *Layer 1*—Signal: objective evidence the buyer already recognises as credible.
- *Layer 2*—Insight: interpretation or context the buyer cannot easily produce internally.
- *Layer 3*—Risk framing: a clear view of what failure looks like, and the consequences of choosing incorrectly.
- *Layer 4*—Economic clarity: an explicit link between cost, outcomes, and time-to-value.

Most traditional sales activity over-indexes on Layer 4. Features, pricing, and commercial terms. High-performing sellers operate across all four, especially the first two, where credibility and differentiation are established early.

---

[1] McKinsey & Company, "The State of AI in Sales," 2023.

Before vs After: The Commercial Cost of Staying Traditional

The difference between traditional selling and AI-augmented selling isn't theoretical. It shows up in measurable commercial outcomes: revenue, margin, cycle time, and forecast quality.

Across mid-market and enterprise teams, the shift typically presents as follows:

Metric	Traditional sales motion	AI-augmented sales motion
Sales cycle length	110–140 days	70–95 days
Late-stage deal loss	25–35%	10–15%
Average discounting	18–25%	8–12%
Forecast accuracy	60–70%	85–95%
Revenue per rep	Baseline	1.3–1.6×

These levers affect NRR/retention indirectly, but they hit valuation directly through cash conversion, gross margin, and predictability. In a market that prices risk, predictability is a premium.

The Sales Archetypes Being Phased Out

Several legacy sales archetypes are losing apply as buying becomes more instrumented:

- The information provider: quickly commoditised when buyers can access credible, comparable information instantly.
- The relationship-only seller: trusted, but rarely decisive once evaluation is structured and cross-functional.

- The intuition-led seller: increasingly overruled by data, benchmarks, and formal risk review.

By contrast, the AI-augmented seller uses technology to reduce noise, improve preparation, and raise judgement at the points that matter.

Harvard[2] Business Review has documented that organisations applying analytics and AI in commercial functions can outperform peers across measures such as productivity and decision quality, including outcomes that map to quota attainment, cycle efficiency, and forecast reliability.[3]

The Sales Relevance Test

Most modern sales engagements now pass through four practical tests:

- Can you surface risks the buyer has not yet accounted for?
- Can you translate data into commercial consequences a CFO will recognise?
- Can you intervene early enough to prevent a deal from stalling or being de-scoped?
- Can you defend price with evidence—rather than persuasion?

Fail one of these and procurement will compensate by tightening terms, expanding competitive pressure, or compressing margin.

Why Resistance to AI Is a Strategic Risk

Resistance to AI is often framed as preserving "human selling." In practice, it usually preserves familiarity. Habits that feel effective because they once were.

---

[2] Harvard Business Review, "How AI Is Changing Sales," 2022

Research from MIT Sloan[3] suggests that well-implemented AI can lift productivity while reducing cognitive load. [4] Sellers who refuse to adapt don't become more human. They become easier to replace.

Throughout this book, I'll introduce you to the QUANTUM Framework. A systematic approach to AI-powered sales that ensures you're drawing on AI capabilities at every stage of your sales process. QUANTUM isn't just a methodology; it's a continuous cycle of improvement that compounds over time, making you progressively more effective with every deal you work

---

[3] MIT Sloan Management Review, "Artificial Intelligence in Revenue Operations," 2023.

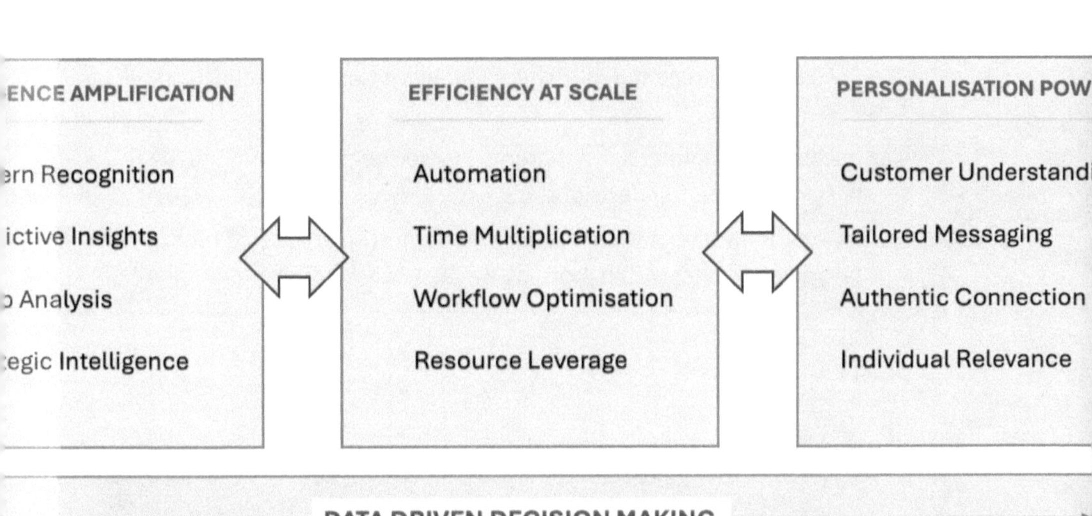

Figure 1.1: The Three Pillars of AI-Powered Sales

CHAPTER 2

THE QUANTUM FRAMEWORK

A New Operating System for Sales

> **By the end of this chapter, you will be able to:**
>
> - Describe QUANTUM as a Sales Operating System (not a one-time methodology) built for AI-era buying behaviour.
> - Explain each stage of QUANTUM and what "good" looks like across the full cycle.
> - Justify why seven stages are required now (AI reduces cognitive load and enables precision).
> - Understand how insights compound across stages and why QUANTUM improves with each iteration.
> - Adopt the core principle: human judgment + machine intelligence, with clear division of labour.

Traditional sales methodologies were built for a different era. SPIN Selling, Challenger Sale, Solution Selling. These frameworks served us well when information was scarce, relationships took months to build, and sales cycles moved at human speed. But we're no longer operating in that world.

Today's buyers are more informed, more sceptical, and more demanding than ever. They've already consumed half your content before you know they exist. They're comparing you to competitors in real-time. They expect personalisation at scale and insights that prove you understand their business better than they do themselves. The old playbooks, no matter how well-executed, simply can't keep pace with these expectations.

At the same time, artificial intelligence has fundamentally changed what's possible in sales. We can now analyse thousands of data points in seconds, predict objections before meetings happen, personalise content for hundreds of prospects simultaneously, and identify patterns that human observation would never catch. The question isn't whether AI will transform sales. It's whether you'll be among those who tap into it or those who get left behind.

This is why I developed the QUANTUM Framework. It's not a minor adjustment to existing methodologies. It's a complete reimagining of the sales process for the AI age. A systematic approach that amplifies human judgment with machine intelligence at every stage.

What is QUANTUM?

QUANTUM is an acronym representing seven interconnected stages that form a complete sales operating system:

Q **- Qualify:** Identify high-potential prospects using AI-powered pattern recognition and predictive analytics

U **- Understand:** Develop deep customer intelligence across organisational, stakeholder, and individual levels

A **- Anticipate:** Predict objections, risks, and opportunities before they materialise in conversations

N **- Navigate:** Orchestrate complex deals through organisational politics and decision-making mazes

T **- Tailor:** Personalise your approach at scale without sacrificing quality or authenticity

U **- Unify:** Integrate systems, data, and insights into a effortless sales ecosystem

M – Measure: Track leading indicators and optimise performance using AI-enhanced analytics

M – Measure: Track leading indicators and optimise performance using AI-enhanced analytics

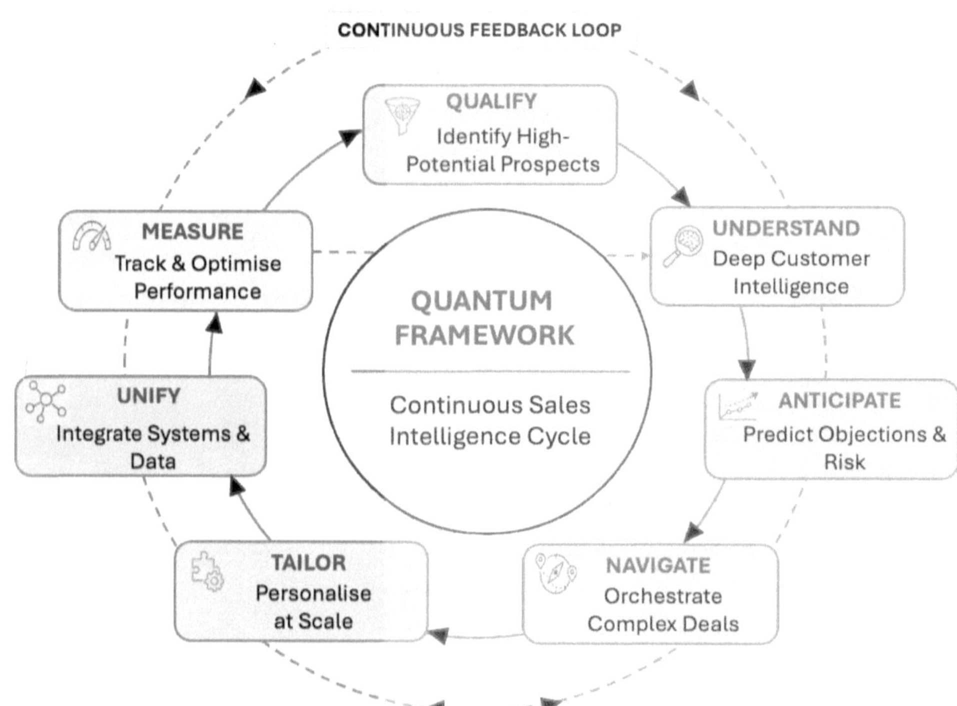

Figure 2.1: *The QUANTUM Framework*
Seven Stages *of Continuous Sales Intelligence*

Each stage builds on the previous one, creating a compounding effect that dramatically amplifies your effectiveness. But here's what makes QUANTUM different from traditional sales frameworks: it's not just a linear process you execute once per deal. It's a continuous cycle where insights from each stage inform and improve all the others.

Why Seven Stages?

You might wonder why we need seven stages when most sales methodologies get by with three or four. The answer is simple: AI enables a level of sophistication that wasn't previously practical.

In the past, we had to simplify sales processes because human cognitive load is limited. You could only track so many variables, remember so many details, and juggle so many activities before quality suffered. So, we created broad stages—"Discovery," "Proposal," "Close"—and accepted that a lot of fine points got lost in those big buckets.

AI removes those constraints. You can now operate with far more precision because the technology handles the complexity while you focus on judgment and relationships. The seven stages of QUANTUM represent the actual distinct phases where different types of intelligence—human and artificial—create maximum leverage.

Think of it this way: a Formula 1 race car has far more instrumentation than a regular car not because it's more complicated for the sake of complexity, but because that precision enables performance that wouldn't otherwise be possible. QUANTUM gives you F1-level precision in your sales process.

How the Framework Works

Let me walk you through how the stages connect and reinforce each other.

Everything starts with **Qualify**. This is where AI helps you escape what I call the "activity trap". That dangerous illusion that more emails, more calls, and more meetings automatically equal more revenue. Instead of chasing every possible opportunity, you use predictive analytics and pattern recognition to identify prospects who match your ideal customer profile and demonstrate genuine buying signals. This isn't about working less; it's about directing your energy where it will produce results.

Once you've identified the right prospects, **Understand** is where you develop deep intelligence about their world. AI-powered research tools can surface insights about organisational dynamics, competitive pressures, and individual motivations that would take traditional methods weeks to uncover. But the key is synthesis. Turning data into actionable intelligence. You're not just learning facts about the company; you're understanding the forces that drive their decisions.

With that understanding in place, **Anticipate** helps you predict what's coming before it happens. AI can analyse patterns from thousands of previous deals to forecast objections, identify risks, and surface opportunities you might otherwise miss. This is where amateurs get blindsided and professionals demonstrate mastery. When you walk into a meeting already prepared for the concerns that haven't been voiced yet, you're operating at a different level entirely.

Guide is about steering complex sales cycles through organisational politics and decision-making mazes. In enterprise sales especially, the org chart rarely reflects actual power structures. AI helps you map stakeholder networks, understand influence patterns, and identify the hidden decision-makers who often kill deals from the shadows. This intelligence allows you to orchestrate multi-threaded strategies that address the real dynamics at play.

Once you understand the field and key players, **Tailor** is where you customise your approach for maximum resonance. Here's the paradox: everyone wants personalisation, but creating truly personalised content and strategies for each prospect seems impossibly time-consuming. AI solves this by generating genuinely customised proposals, presentations, and communications at scale while you maintain quality control and strategic oversight. You get the impact of bespoke work with the efficiency of systematised processes.

Unify addresses one of the biggest headaches in modern sales: the explosion of tools, data sources, and communication channels. Without integration, you end up with information trapped in silos, duplicate work, and important signals falling through the cracks. This stage is about creating an ecosystem where insights flow cleanly between systems, eliminating friction and ensuring nothing gets missed. When your tools work together intelligently, you spend less time on administrative work and more time selling.

Finally, **Measure** closes the loop by tracking what predicts success. This goes far beyond traditional sales metrics like revenue and close rates. AI analytics help you identify leading indicators. The early signals that forecast outcomes before they happen. You learn which activities move deals forward, which prospects are likely to convert, and where to invest your time for maximum return. These insights then feed back into the Qualify stage, continuously improving your targeting and approach.

The Continuous Cycle

Here's the crucial insight: QUANTUM isn't a one-way street you travel once per deal. It's a cycle that continuously refines itself.

Every deal you work teaches the system something new. When you Measure the results of your approach, those insights improve how you Qualify prospects. When you Guide complex stakeholder dynamics, that experience enhances your ability to Understand similar situations later. When you Anticipate objections that don't materialise, the system learns to adjust its predictions. The framework gets smarter every time you use it.

Taken together, this creates a compounding advantage. While your competitors are executing the same playbook again, you're operating a system that improves with every iteration. Six months into using QUANTUM, you're not just doing the same things faster. You're doing fundamentally better work because the framework has learned from hundreds of interactions.

Human + Machine: The Core Principle

Before we dive into each stage in the following chapters, I want to be crystal clear about something: QUANTUM is not about replacing human judgment with artificial intelligence. It's about combining both to achieve what neither could accomplish alone.

AI excels at pattern recognition, data processing, prediction, and scale. Humans excel at judgment, creativity, relationship building, and working through ambiguity. QUANTUM works by having each do what they do best.

At its simplest, the framework automates the automatable—research, data analysis, content generation, risk assessment—so you can focus on what requires human touch: building trust, understanding subtle contexts, making strategic decisions, and creating genuine connections with customers.

Every stage of QUANTUM includes specific guidance on where to capitalise on AI and where to apply human judgment. You'll never be asked to blindly follow algorithmic recommendations or remove yourself from critical decisions. Instead, you'll learn to interpret AI insights, apply contextual understanding, and make informed choices that blend machine intelligence with human wisdom.

What You'll Gain

If you implement QUANTUM systematically, here's what changes:

Your targeting becomes surgical. Instead of spraying your message to anyone who might possibly buy, you focus on prospects who match your ideal profile and demonstrate actual buying signals. Your qualification accuracy improves dramatically, which means higher conversion rates and shorter sales cycles.

Your customer understanding deepens exponentially. You'll walk into conversations knowing more about the prospect's business, competitive arena, and internal dynamics than many of their own employees. This isn't about showing off; it's about offering insights that genuinely help them make better decisions.

You stop being surprised. Objections you anticipated. Risks you saw coming. Opportunities you positioned for. Instead of constantly reacting, you're three steps ahead, which completely changes the dynamic of customer conversations.

Your personalisation scales. You can deliver genuinely customised experiences to dozens or hundreds of prospects without sacrificing quality. Each person feels like you've crafted an approach specifically for them. Because you have, with AI handling the execution at scale.

Your deals move faster. When you know who to talk to, what to say, and when to say it, friction disappears. Decision-makers get the information they need when they need it. Objections get addressed before they become blockers. Deals that might have dragged for months close in weeks.

Your forecast accuracy improves. You're no longer guessing which deals will close based on gut feel and hopeful thinking. AI-powered leading indicators tell you which opportunities are real and which are pipe dreams, allowing you to forecast with confidence and allocate resources intelligently.

You work smarter, not just harder. This might be the most important benefit. QUANTUM isn't about adding more to your plate. It's about removing low-value activities so you can focus on high-impact work. You spend less time on research, administration, and qualification, and more time selling.

The Road Ahead

The next seven chapters will take you through each stage of QUANTUM in detail. You'll get specific techniques, real-world examples, and practical implementation guidance. I'll show you exactly which tools work best for each stage, how to avoid common pitfalls, and how to customise the framework for your specific situation.

But here's what I want you to understand before we dive in: QUANTUM isn't a theoretical model I came up with in isolation. Every technique, every recommendation, every example comes from real experience. Mine and that of the sales teams I've built and coached. I've used this framework to close seven-figure enterprise deals, build predictable pipelines in chaotic markets, and consistently hit targets when others were struggling.

More significantly, I've seen it work across different industries, company sizes, and sales contexts. Whether you're a solo consultant selling professional services or an enterprise sales professional managing complex organisational sale, the principles remain the same. The specific tools and tactics might vary, but the framework holds.

A Note on Implementation

You might be thinking: "This sounds great, but how do I actually implement all of this while carrying a full quota?"

Fair question. Here's the good news: you don't have to adopt the entire framework at once. QUANTUM is modular. You can start with one stage. Typically Qualify or Measure, since those often provide the quickest wins. And progressively layer in additional stages as you build competence and see results.

Each chapter includes an implementation section with specific first steps, tool recommendations, and success metrics. I've designed these to be practical and immediately actionable. You could read Chapter 3 on Monday and start improving your customer intelligence work on Tuesday.

That said, the real power of QUANTUM comes from integration. When all seven stages work together, reinforcing and informing each other, the results compound exponentially. So, while starting with one stage makes sense, keep the complete framework in mind as your destination.

Why This Matters Now

We're at an inflection point in sales. The gap between those who draw on AI effectively and those who don't is growing exponentially. It's no longer a marginal advantage. It's the difference between thriving and struggling.

I've watched talented salespeople burn out trying to keep up with demands that human effort alone can't meet. I've seen deals lost not because the product wasn't right or the price wasn't competitive, but because the competitor used AI to understand the customer better, move faster, and deliver more personalised value.

The good news is that the tools exist today to change this equation. AI that seemed like science fiction three years ago is now accessible, affordable, and practical. The barrier isn't technology anymore. It's knowledge and methodology. That's what QUANTUM provides: a systematic way to use these capabilities and translate them into sales results.

Moving Forward

In Chapter 3, we'll dive into the first stage: Qualify. You'll learn exactly how to use AI to identify high-potential prospects, build ideal customer profiles that predict success, and escape the activity trap that exhausts most salespeople while producing mediocre results.

But before you turn the page, take a moment to consider where you are right now in your sales approach. How much time do you spend on prospects who were never going to buy? How often are you surprised by objections you should have seen coming? How much valuable customer intelligence is trapped in different systems or never captured at all?

QUANTUM is designed to solve these problems systematically. Not with motivational platitudes or surface-level tips, but with a complete framework that combines human judgment with AI capabilities to produce consistent, predictable, and scalable results. Let's get to work.

CHAPTER OUTCOME

☐ A one-sentence definition of QUANTUM as your sales operating system

☐ The 7 stages written out with your own "what good looks like" for each stage

☐ A simple note on how insights compound across stages (where the flywheel effect comes from)

☐ A decision on where AI helps most in your process (and where humans must stay in control)

☐ A baseline maturity score (1–5) for your current capability in each QUANTUM stage

☐ One "fast win" stage to improve first (based on impact + feasibility)

☐ A plan for how you'll run the framework weekly (cadence + review habit)

CHAPTER 3

Qualify – Finding Your Ideal Prospects

By the end of this chapter, you will be able to:

- Escape the "activity trap" and define what qualified truly means (Fit, Need, Intent).
- Build a practical ICP using AI to identify patterns and define inclusion/exclusion criteria.
- Implement AI-powered qualification methods (predictive scoring, intent data, disqualification signals).
- Run a qualification conversation using hypotheses, truth-revealing questions, and AI validation.
- Prioritise pipeline using a Qualification Matrix and apply disqualification as a discipline.

Escaping the activity trap and focusing on deals you can win

Early in my sales career, I made the mistake most new salespeople make I equated activity with progress. More calls meant more opportunities. More meetings meant more pipeline. More emails meant more chances to close deals. I was working 60-hour weeks, constantly busy, and consistently missing quota.

My manager finally sat me down and asked a simple question: "Of your last 20 deals, how many actually closed?"

I did the math. Four. A 20% close rate.

"Now think about those four deals that closed. When you first talked to them, how did you know they were real opportunities?"

I thought about it. They all had clear pain points. They all had budget. They all had a defined timeline. They all fit our ideal customer profile almost perfectly.

"Now think about the 16 that didn't close. How many of those were never really qualified in the first place?"

In real terms, the answer was uncomfortable: at least 12 of them. I'd spent months working deals with prospects who didn't have budget, didn't have authority, didn't have a real problem we solved, or weren't in our sweet spot. I was busy, but I was busy with the wrong things.

That conversation changed my career. I learned to qualify ruthlessly. To spend my time on prospects who matched our ideal profile and showed genuine buying signals, and to disqualify everything else quickly so I could focus where I could win.

This is the Qualify stage of the QUANTUM Framework. The foundation that everything else builds on. If you're focused on the wrong prospects, nothing else matters. Your deep understanding, your anticipation, your navigation, your personalisation. All wasted on deals you were never going to close.

Qualification isn't about working less. It's about working smarter by directing your finite time and energy toward opportunities where you can create value and win deals.

The Activity Trap

Most sales organisations are caught in what I call the "activity trap". The mistaken belief that more activity automatically leads to more results.

This shows up in several ways:

Volume metrics over outcome metrics: Sales managers measure calls made, emails sent, meetings booked. These are easy to count and easy to manage. But they're lagging indicators at best, and misleading ones at worst. A rep making 100 calls to poorly qualified prospects will perform worse than one making 20 calls to well-qualified ones.

Fear of disqualification: Reps are afraid to disqualify opportunities because it makes their pipeline look smaller. A pipeline with 30 mediocre opportunities feels more impressive than one with 10 excellent ones. But the second pipeline will close more revenue.

Unclear qualification criteria: Many organisations have vague qualification frameworks (BANT, MEDDIC, etc.) that everyone claims to use but no one applies consistently. When qualification is subjective, reps qualify optimistically, keeping marginal deals in the pipeline far too long.

Reactive prospecting: Reps chase whatever leads come in rather than strategically targeting prospects most likely to convert. Marketing sends a lead, they work it. Someone fills out a form, they follow up. They're reacting to inbound rather than proactively pursuing ideal customers.

The result: sales reps spend most of their time on prospects who will never buy, while under-investing in prospects who would buy if properly engaged. They're exhausted, their forecasts are unreliable, and their close rates are terrible.

AI and modern sales intelligence change this equation completely. Instead of guessing which prospects are worth pursuing, you can identify patterns that predict conversion. Instead of optimistically keeping marginal deals alive, you can objectively assess deal quality. Instead of reacting to whatever comes in, you can proactively target prospects that match proven success patterns.

What Does "Qualified" Actually Mean?

Before we talk about how to qualify effectively, let's define what we mean by a qualified prospect. Different organisations have different criteria, but at a fundamental level, a qualified prospect has three characteristics:

1. Fit: They Match Your Ideal Customer Profile

Fit is about whether this prospect is the type of customer you're designed to serve successfully. This includes:

Firmographic fit:

- Company size (revenue, employees)
- Industry and vertical
- Geographic location
- Company stage (startup, growth, enterprise, etc.)

Technographic fit:

- Current technology stack
- Technical sophistication
- Integration requirements
- Platform preferences

Situational fit:

- Current challenges that your solution addresses
- Organisational maturity for the change you require
- Budget availability in your price range
- Decision-making structure you can work through

Fit isn't about whether you could technically work with this company. It's about whether you can deliver exceptional value based on how your solution is designed and what you're best at.

Example: If you've built a sales automation platform optimised for fast-growing B2B tech companies with 50-200 employees, you might be able to sell to a 10,000-person manufacturing company. But you won't be able to deliver the same value, and you'll struggle to compete with solutions designed for that market.

2. Need: They Have a Problem You Can Solve

Need means they have a genuine problem or opportunity that your solution addresses, and that problem is significant enough to justify action.

Pain intensity: Is this a "nice to have" or a "must solve"? Are they frustrated enough with the status quo to change? Do they feel urgency, or is this just exploratory?

Problem alignment: Is their problem what your solution solves? Not what you can stretch your solution to address, but what you genuinely solve well? Prospects sometimes have problems adjacent to what you solve, but if you're not addressing their core issue, you won't create compelling value.

Timing readiness: Is this a problem they're ready to solve now, or something they're thinking about for next year? Timing determines whether this is an active opportunity or something to nurture for later.

Example: A prospect might fit your ICP perfectly (right size company, right industry) but if they're happy with their current solution, have no pain points driving change, and aren't facing any external forcing function (like contract renewal or regulatory requirement), there's no real need. You could convince them your solution is better, but without genuine pain, they won't prioritise the change.

3. Intent: They're Actually in-Market and Ready to Evaluate

Intent is about whether they're actively looking for a solution or just passively gathering information. This is the difference between a real opportunity and a timewaster.

Active evaluation signals:

- They've reached out to multiple vendors
- They have a defined timeline
- They've allocated budget or gotten approval to spend
- Multiple stakeholders are involved in conversations
- They're asking detailed questions about implementation and pricing

Passive exploration signals:

- They're "just looking" or "gathering information"
- Timeline is vague or TBD
- No budget has been allocated yet
- Only one person is engaged, with no broader organisational involvement
- Questions are high-level and theoretical

The challenge: passive explorers can turn into active buyers. The key is recognising which stage they're in and investing accordingly. You nurture passive explorers efficiently until they show real intent, rather than treating them like active opportunities.

Building Your Ideal Customer Profile (ICP)

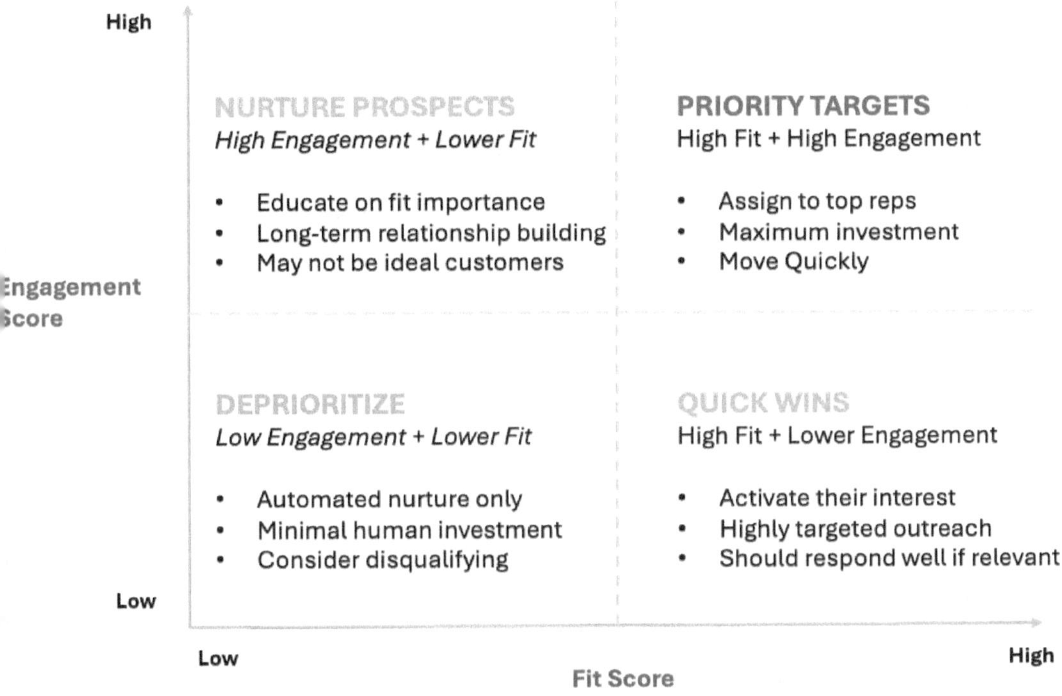

Figure 3.1: ICP Analysis Matrix for Pipeline Prioritisation

Most companies have vague ICPs: "Mid-market B2B software companies." That's too broad to be actionable. An effective ICP is specific enough to immediately include or exclude most prospects.

Here's how to build one:

Step 1: Analyse Your Best Customers

Start with your most successful customers. Not just those who spend the most, but those who:

- Bought quickly with minimal friction
- Achieved strong results with your solution
- Stayed with you long-term with low churn risk
- Expanded usage over time
- Advocate for you and provide references

For each of these best customers, document:

Firmographic characteristics:

- Revenue range
- Employee count
- Industry/vertical
- Geographic location
- Company age and stage
- Ownership (private, PE-backed, public, etc.)

Technographic characteristics:

- Core technology platforms they use
- Technology spending patterns
- Technical sophistication level
- IT structure and decision-making

Behavioural characteristics:

- How they found you (referral, inbound, outbound, etc.)

- Sales cycle length
- Deal size
- Stakeholders involved in decision
- Key pain points they had when they bought

Outcome characteristics:
- How fast they achieved value
- Key use cases they implemented
- Expansion patterns
- Retention and satisfaction metrics

Step 2: Identify Patterns with AI

Once you've documented 20-30 of your best customers, AI can identify patterns that aren't obvious from manual analysis.

Tools like Salesforce Einstein, HubSpot Predictive Lead Scoring, or specialized platforms like MadKudu and Infer analyse your customer data and identify which characteristics most strongly correlate with good outcomes.

AI might reveal patterns like:
- "Companies with 75-150 employees close 2.3x faster than those with 151-300"
- "Prospects using Salesforce as their CRM have 68% higher close rates than those using HubSpot"
- "Companies in healthcare and financial services have identical close rates but financial services companies expand 3x more over two years"
- "When you engage both VP of Sales and VP of Marketing early, close rate is 47%; when only VP of Sales, it's 22%"

These insights let you refine your ICP beyond gut feel to data-driven targeting. You might discover your assumed ICP is wrong. You thought you were best for mid-market, but data shows you win more and deliver better results in small enterprise.

Step 3: Define Clear Inclusion/Exclusion Criteria

Based on this analysis, create specific qualification criteria:

Must-Have Criteria (non-negotiable for qualification):

- Industry: B2B SaaS or professional services
- Size: $10M-$100M revenue OR 100-500 employees
- Technology: Using Salesforce, HubSpot, or Microsoft Dynamics
- Structure: Has dedicated sales operations function
- Pain point: Struggling with sales forecast accuracy or pipeline visibility

Strong Fit Criteria (ideal but not required):

- Growth rate: 20%+ YoY revenue growth
- Sales team: 15-50 quota-carrying reps
- Technology spends: Willing to invest $50K-$250K in sales tools
- Geography: North America or Western Europe (for timezone/support reasons)
- Advocate: Has internal champion who can drive consensus

Disqualifying Criteria (immediate disqualification):

- Company size: <50 employees or >2,000 employees
- Industry: Healthcare (regulatory complexity we can't support), manufacturing (not our use case)
- Technology: Homegrown CRM (integration too complex)
- Decision-making: Requires 6+ month procurement process
- Budget: Can't invest at least $30K annually

Notice how this clarity makes qualification immediate. When you encounter a prospect, you can quickly determine whether they're worth pursuing or should be disqualified (or nurtured until circumstances change).

Step 4: Create a Scoring Model

Beyond binary qualification (yes/no), create a scoring system that prioritises opportunities.

Assign point values to different characteristics:

Firmographic scoring:

- 10-50M revenue: +10 points
- 50-100M revenue: +15 points
- B2B SaaS: +15 points
- Professional services: +10 points
- 100-500 employees: +10 points

Technographic scoring:

- Using Salesforce: +15 points
- Using HubSpot: +10 points
- Using Outreach/SalesLoft: +5 points

Behavioural scoring:

- Inbound inquiry: +10 points
- Downloaded multiple assets: +5 points
- Attended webinar: +5 points
- Engaged multiple stakeholders: +15 points
- Mentioned timeline: +10 points

Total score determines priority: 70+ points = high priority, 50-69 = medium priority, <50 = low priority or disqualify.

AI can automate this scoring, instantly evaluating every new lead and routing high-priority prospect to top reps while lower-priority prospects get automated nurture sequences.

AI-Powered Qualification: Beyond Traditional Frameworks

Traditional qualification frameworks like BANT (Budget, Authority, Need, Timeline) or MEDDIC (Metrics, Economic Buyer, Decision Criteria, Decision Process, Identify Pain, Champion) are useful but limited. They rely on asking questions and trusting prospects to give accurate answers.

AI-powered qualification is fundamentally different. It doesn't just rely on what prospects tell you. It looks at patterns across thousands of deals to predict which prospects will convert, even when the prospect themselves doesn't know yet.

Predictive Lead Scoring

Modern CRM systems and sales intelligence platforms use machine learning to score leads based on how similar they are to past deals that closed.

The AI considers:

- Firmographic match to your ICP
- Behavioural signals (website visits, content engagement, email responses)
- Fit indicators (technology stack, org structure, company stage)
- Engagement patterns (response time, meeting acceptance, stakeholder involvement)
- External signals (funding events, leadership changes, tech stack changes)

Instead of manually evaluating each lead, AI instantly scores every prospect: "Based on similarity to your best customers and their current behaviour, this lead has an 81% probability of converting if engaged within 48 hours."

This allows you to prioritise automatically. Your best reps work highest-probability opportunities, while lower-probability leads get automated or junior rep engagement.

Example: Clari's AI analyses over 2 million data points per opportunity, comparing each new lead to historical patterns. It can predict with remarkable accuracy which deals will close, which will stall, and which will be lost. Often before the sales rep recognises it themselves.

Intent Data and Buying Signals

Traditional qualification asks prospects, "Are you actively looking?" AI doesn't need to ask. It monitors behaviour to detect genuine buying intent.

Intent data platforms (like Bombora, TechTarget, G2) track content consumption across the web. When multiple people from a company research topic related to your category—reading reviews, downloading whitepapers, watching demos—that indicates active evaluation even if they haven't contacted you yet.

AI synthesises this intent data with other signals:

- Website behaviour: Which pages they visit, how long they stay, how often they return
- Email engagement: Open rates, click rates, reply to patterns
- Social media: Who follows you, who engages with your content, what they share
- Technology changes: When they adopt adjacent technologies, it often signals readiness for yours
- Organisational changes: New executives, new funding, new products launch. All create buying windows

Instead of waiting for prospects to raise their hand, you can proactively reach out when signals indicate they're in-market: "I noticed your team has been researching sales automation platforms based on recent activity, and you just hired a new VP of Sales. Most companies in similar situations are evaluating tools like ours within 60-90 days of that hire. Would it make sense to have a conversation now while you're still early in your evaluation?"

This transforms qualification from reactive (evaluating inbound leads) to proactive (identifying prospects showing buying intent before they reach out).

Pattern-Based Disqualification

Just as AI can identify which prospects are likely to buy, it can also identify which ones are likely to waste your time. Even when they seem interested.

Red flag patterns AI can spot:

- "Prospects who engage heavily for 2 weeks then ghost for 30+ days rarely re-engage" (so disqualify rather than keep chasing)
- "When only one stakeholder is engaged after 3 touchpoints, close rate drops to 8%" (so either expand stakeholders or disqualify)
- "Companies that haven't allocated budget within 60 days of initial contact have 89% probability of never allocating budget" (so disqualify or move to long-term nurture)
- "Prospects who ask for proposals without first having a discovery call close at 3% rate" (so don't waste time on proposals, insist on discovery first)

On closer inspection, thise patterns let you disqualify confidently even when a prospect seems interested. If their behaviour matches patterns that historically don't convert, you're not being pessimistic. You're being data-driven.

Real-Time Qualification Insights During Conversations

most advanced application of AI in qualification happens during live sales conversations. Conversation intelligence platforms like Gong and Chorus.ai analyse calls in real-time and can surface:

"This prospect just mentioned they 'need to think about timing'. Historically, this phrase in discovery calls predicts 76% probability of stalled deal. Recommend addressing timeline explicitly before ending call."

"Prospect has mentioned 'budget' three times but hasn't specified an amount. Historical patterns suggest this indicates budget uncertainty. Recommended question: 'What budget range has been allocated for this initiative?'"

"This is the third call where only the VP of Marketing has joined. In similar deals, when economic buyer hasn't engaged by call 3, close rate drops to 12%. Recommend asking: 'Who else should be involved in this evaluation?'"

Instead of realising qualification issues after the call, you can address them during the conversation while you still have the prospect's attention.

The Qualification Conversation

Even with all this AI-powered intelligence, you still need to have actual qualification conversations with prospects. The difference is that AI tells you what to validate and what questions to prioritise.

Here's how to structure effective qualification discussions:

Start with Hypotheses, Not Interrogation

Traditional discovery feels like an interrogation: "What's your budget? Who's the decision-maker? What's your timeline? What challenges are you facing?"

Better approach: Come in with hypotheses based on AI insights and validate them conversationally.

Instead of: "What challenges are you facing with your current sales process?"

Try: "Most companies your size in SaaS is struggling with forecast accuracy as they scale. Visibility into pipeline becomes really difficult once you hit about 30 reps. Is that consistent with what you're experiencing, or are your challenges different?"

This demonstrates you understand their world, makes the conversation collaborative rather than interrogative, and quickly confirms or refutes your qualification hypothesis.

Ask Questions That Reveal Truth, Not Answers Prospects Think You Want

Prospects often tell you what they think you want to hear. They claim to have budget when it's not approved. They say they're ready to move forward when they're just exploring. They identify themselves as decision-makers when they're not.

Ask questions that reveal truth through their answers:

To assess budget reality:

- Not: "Do you have budget?" (Everyone says yes)
- Instead: "Walk me through your budget allocation process. When was this budget approved? What other initiatives are competing for it?"

To assess decision authority:

- Not: "Are you the decision-maker?" (Everyone says yes or "I'm the key person")
- Instead: "Tell me about the last significant technology purchase you made. Who was involved? How did the decision process work?"

To assess urgency:

- Not: "When do you want to implement?" (They'll give you optimistic timelines)
- Instead: "What happens if you don't solve this problem in the next 90 days? What's the cost to the business?"

To assess need intensity:

- Not: "How important is this problem?" (Everything is "very important")
- Instead: "What have you already tried to solve this? What stopped you from fixing it before now?"

The answers reveal whether this is a real opportunity or wishful thinking.

Use AI to Validate What You Hear

After qualification conversations, use AI to check whether what you heard aligns with patterns:

Prospect claims: "We have budget approved and are ready to move quickly."

AI insight: "This prospect's company has procurement processes that historically take 4-6 months. Timeline is likely optimistic. Recommend validating with procurement involvement."

Prospect claims: "I'm driving this decision."

AI insight: "Job title is Director-level. In this company size, Director-level contacts close deals 18% of the time without VP or C-level involvement. Recommend identifying economic buyer."

This doesn't mean prospects are lying. They often believe what they're telling you. AI provides reality-check based on what happens in similar situations.

The Qualification Matrix: Prioritising Your Pipeline

Not all qualified opportunities are equal. You need a systematic way to prioritise where you invest time.

Use a simple 2x2 matrix:

X-axis: Fit Score (How well do they match your ICP?)

- Low = Weak fit, many gaps from ideal profile
- High = Strong fit, matches ICP closely

Y-axis: Engagement Score (How much buying intent and stakeholder involvement?)

- Low = Limited engagement, few buying signals, single-threaded
- High = Strong engagement, multiple buying signals, stakeholders involved

This creates four quadrants:

Top Right: Priority Targets (High Fit + High Engagement)

- These are your best opportunities
- Assign to your best reps
- Invest maximum time and resources
- Move quickly—competitors see the same signals

Top Left: Nurture Prospects (High Engagement + Lower Fit)

- They're interested but not ideal customers
- Risk: You might win but won't deliver strong value
- Strategy: Educate them on why fit matters, refer to better alternatives if appropriate, or accept lower win priority

Bottom Right: Quick Wins (High Fit + Lower Engagement)

- Perfect profile but not actively engaged yet
- Strategy: Invest in activating their interest. They should respond well if you're relevant
- Use personalisation and relevant outreach to increase engagement

Bottom Left: Deprioritise (Low Fit + Low Engagement)

- Neither good fit nor showing interest

- Strategy: Automated nurture at most, don't invest human time
- Be comfortable disqualifying entirely

Most sales reps don't consciously prioritise this way. They treat all "opportunities" equally, spreading their time thin rather than concentrating it where they have the highest probability of success.

AI can automatically plot every opportunity on this matrix and route them accordingly. Priority Targets to senior reps for immediate attention, Quick Wins to specialized teams focused on activation, Nurture Prospects to automated sequences with occasional human touchpoints, and Deprioritise prospects filtered out entirely.

Qualification as Ongoing Process

Qualification isn't a one-time checkpoint. Deals change. What seemed like a Priority Target can become Deprioritise if engagement drops or you learn information that reveals fit problems.

Continuously re-qualify opportunities by monitoring:

Engagement trends: Is stakeholder involvement increasing or decreasing? Are they responding faster or slower? Are they consuming more content or less?

Competitive dynamics: Are other vendors getting more involved? Is the prospect comparing you to alternatives you can't compete with?

Organisational changes: Did your champion leave? Did budget get reallocated? Did priorities shift?

Timeline slippage: Are deadlines getting pushed? Is "next month" turning into "next quarter"?

AI can monitor these signals automatically and alert you when deal health deteriorates: "Acme Corp engagement score dropped 34% in the past 2 weeks. Champion hasn't responded to last 3 emails. Last meeting was rescheduled twice. Recommend re-qualifying this opportunity."

As a result, this saves you from clinging to dead deals because you've invested time in them. Sunk cost fallacy kills sales productivity. Better to recognise when an opportunity isn't real and reallocate time to genuine prospects.

The Discipline of Disqualification

The hardest part of qualification isn't identifying good prospects. It's disqualifying bad ones. Reps resist disqualification because:

- It makes their pipeline smaller (even though the smaller pipeline is more likely to convert)
- It feels like giving up (even though it's redirecting effort to better opportunities)
- They're optimistic by nature (even though optimism without data is just hope)
- They're measured on pipeline size (even though pipeline quality matters more)

Here's how to build disqualification discipline:

Set Clear Disqualification Criteria

Don't make disqualification a judgment call. Have specific, objective criteria:

- After 3 touchpoints with no stakeholder expansion → Disqualify
- After 60 days with no budget discussion → Disqualify to long-term nurture
- After 2 rescheduled meetings → Disqualify unless champion provides specific reason and commitment
- When key pain point doesn't match what you solve → Disqualify immediately

When criteria are objective, disqualification becomes a process, not a judgment of the rep's abilities.

Celebrate Disqualification

Organisations that excel at qualification celebrate when reps disqualify appropriately. "Sarah disqualified 5 opportunities this week that were wasting her time, and she's now focused on 3 high-probability deals. That's exactly what we want."

Measure and reward qualification quality, not just pipeline quantity:

- Track close rate (high close rate indicates good qualification)
- Track cycle time (well-qualified deals close faster)
- Track time spent per opportunity (disqualifying frees up time for better prospects)

Use AI to Suggest Disqualification

When AI recommends disqualification based on patterns, it removes the emotional element. It's not the rep giving up. Its data indicating this opportunity doesn't match success patterns.

"This opportunity has been in discovery for 90 days with no progress to demo stage. Historically, deals that don't advance to demo within 45 days have a 4% close rate. Recommend disqualifying and reallocating time to higher-probability opportunities."

Critically, the rep can override if they have specific information AI doesn't see, but the default should be following data-driven recommendations.

From Qualification to Systematic Targeting

The ultimate evolution of qualification is moving from reactive (qualifying what comes to you) to proactive (systematically targeting prospects that match your success patterns).

Proactive targeting means:

- **Building target account lists** based on ICP rather than waiting for inbound

- **Monitoring for buying signals** across those accounts using intent data
- **Reaching out proactively** when signals indicate evaluation is starting
- **Personalising outreach** based on fit indicators and intent signals

This flips the traditional model. Instead of marketing generating leads that sales qualify, sales intelligence identifies ideal prospects showing buying intent, and sales reaches out with relevant, personalised messaging before competitors even know these prospects are in-market.

Companies like ZoomInfo, Apollo, and Clay enable this by combining:

- Firmographic data (to identify ICP matches)
- Technographic data (to see what tools they use)
- Intent data (to detect buying signals)
- Contact data (to reach the right stakeholders)
- Enrichment data (to personalise outreach)

AI orchestrates this: "These 15 accounts match your ICP perfectly, are showing intent signals in your category, and have new VPs of Sales who typically make buying decisions in their first 90 days. Recommend outbound sequences starting this week."

Putting Qualification into Practice

Let's bring this together with a practical example of how AI-powered qualification works:

Monday morning: AI analyses your pipeline and highlights 3 opportunities:

Opportunity 1: TechStart Inc.

- Fit score: 92/100 (perfect ICP match)
- Engagement: 88/100 (high intent signals)

- Alert: "Multiple stakeholders researching your category, new VP of Sales hired 30 days ago, currently using competitor X whose contract expires in 60 days"
- Recommendation: Priority Target—schedule discovery call this week

Opportunity 2: MegaCorp Global

- Fit score: 45/100 (too large, wrong industry)
- Engagement: 72/100 (they're interested but...)
- Alert: "Three discovery calls but no decision-maker involvement, enterprise procurement process typically 9-12 months"
- Recommendation: Disqualify—wrong fit, too complex, low probability

Opportunity 3: GrowthCo

- Fit score: 85/100 (good fit)
- Engagement: 34/100 (single-threaded, slow responses)
- Alert: "Champion hasn't responded in 3 weeks, no other stakeholders engaged, timeline has slipped twice"
- Recommendation: Move to nurture—not dead but not active, automated touchpoints only

You take the recommendation: focus immediately on TechStart (schedule call, prepare personalised demo, engage multiple stakeholders), disqualify MegaCorp (politely explain fit issues, refer them to enterprise-focused alternative), and move GrowthCo to automated nurture (occasional touchpoints, but no active time investment).

Result: Your pipeline shrunk from 30 opportunities to 15, but your close rate doubled and your cycle time decreased by 35% because you're focused on deals you can win.

The Qualification Advantage

Here's what changes when you qualify systematically with AI:

- **Your close rate improves** because you're working opportunities that match success patterns, not wishful thinking.
- **Your cycle time decreases** because well-qualified deals don't stall. They have real need, real budget, and real urgency.
- **Your forecast accuracy increases** because your pipeline reflects real opportunities, not hopeful prospects.
- **Your stress decreases** because you're not constantly firefighting deals that were never going to close.
- **Your efficiency increases** because you're not wasting time on prospects who don't fit or aren't ready.

Most significantly, you shift from reacting to whatever comes your way to strategically focusing on opportunities where you can create genuine value and win consistently.

In the next chapter, we'll explore Understand—how to develop deep intelligence about the prospects you've qualified so that every conversation demonstrates you genuinely know their business, their challenges, and their motivations.

Because qualifying the right prospects is only the start. Understanding them deeply is what turns opportunities into wins. In the next chapter, we'll explore how to develop that deep understanding systematically. Moving beyond surface-level research to genuine customer intelligence that transforms your sales conversations.

CHAPTER OUTCOME

☐ A defined ICP with clear inclusion + exclusion criteria (not vague descriptors)

☐ A qualification scorecard that covers Fit, Need, Intent (or your chosen equivalent)

☐ A disqualification list (red flags you will actively walk away from)

☐ 10 truth-revealing qualification questions you can use in calls

☐ A shortlist of data sources/signals you'll use for qualification (CRM, intent, firmographic, technographic, etc.)

☐ A simple pipeline prioritisation rule (who gets time first and why)

☐ A process to keep qualification honest (how you validate assumptions)

CHAPTER 4

Understand – Deep Customer Intelligence

By the end of this chapter, you will be able to:

- Build customer understanding across three layers: organisational context, stakeholder dynamics, and individual motivations.
- Apply decision psychology basics (biases and heuristics) to how buyers actually choose.
- Convert research into actionable intelligence using a repeatable AI research workflow (aggregate → synthesise → signal → recommend).
- Translate intelligence into better conversations (moving from generic discovery to insight-led engagement).
- Avoid common understanding errors: confusing data for insight, assuming without validating, overwhelming prospects, and failing to act.

Moving beyond surface-level research to genuine insight

A few years ago, I sat in on a deal review where a sales rep confidently declared she "knew the account inside and out." She'd done her homework: researched the company website, read recent press releases, reviewed the executive team on LinkedIn, and even found their last earnings call transcript. On paper, she was well-prepared.

Then the sales manager asked a simple question: "Why are they actually looking at our solution right now?"

The rep paused. "Well, they're growing fast and need to scale their operations..."

"But they've been growing fast for three years. Why now? What changed?"

Another pause. "I think... their VP of Sales just joined from a company that used our platform?"

"You think, or you know?"

In practice, the rep didn't know. Despite all her research, she'd gathered facts without understanding context. She knew what the company did, but not why they'd act. She knew who the decision-makers were, but not what motivated them individually. She had data, but not intelligence.

This is the gap the Understand stage of the QUANTUM Framework addresses. In the Qualify stage, you identified the right prospects to pursue. Now you need to develop the deep intelligence that turns generic sales conversations into strategic partnerships. You need to move from knowing about customers to genuinely understanding them. Their business pressures, their organisational dynamics, their personal motivations, and the cognitive patterns that drive their decisions.

Recognise that this chapter is about building that understanding systematically, using AI to surface insights that would take months of traditional research, and synthesising information into actionable intelligence that transforms how you sell.

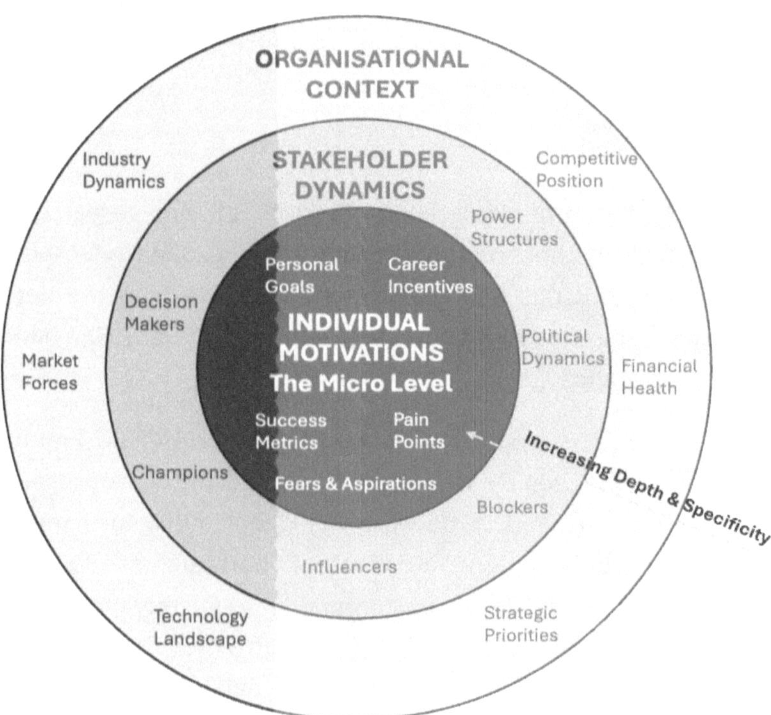

*Figure 4.1: **Three** Layers of Customer Understanding -*
***Complete Customer Intelli**gence Requires Depth Across All Three Layers*

Understanding a customer isn't a single thing. It operates at multiple levels simultaneously. Most salespeople stop at the surface layer, which is why their "research" doesn't translate into better conversations or won deals. Elite sellers—and AI-enhanced sales organisations—develop understanding across three distinct layers:

Layer 1: Organisational Context

This is the macro level. Understanding the company as an entity operating within a competitive terrain, facing specific market pressures, and pursuing defined strategic objectives.

What you need to understand:

Industry dynamics: What forces are reshaping their industry? Is it consolidation, disruption, regulation, technological change? A healthcare company facing new compliance requirements has different priorities than one dealing with private equity consolidation. Understanding which forces are dominant tells you what keeps executives up at night.

Competitive position: Where does this company sit relative to competitors? Are they the market leader defending position, a fast-follower trying to catch up, or a disruptor trying to rewrite the rules? Your value proposition changes dramatically based on their competitive stance. Leaders need to protect margins and operational efficiency. Challengers need to move faster and take risks. Disruptors need to prove their model works at scale.

Financial health and constraints: Are they profitable and growing, or burning cash to gain market share? Are they preparing for IPO, recently acquired, or facing activist investors? Financial context determines what kinds of investments they can make and how they'll evaluate ROI. A company with tight margins scrutinizes every dollar. A well-funded growth company might prioritise speed over cost.

Strategic priorities: What are they trying to accomplish in the next 12-18 months? Not their generic mission statement. Their real priorities. Are they expanding into new markets, launching new products, improving operational efficiency, or dealing with executive transitions? Your solution needs to map to these priorities, not just generic pain points.

Technology field: What systems and platforms are they already using? Not just in your category, but across their entire tech stack. This tells you about integration requirements, change management capacity, and technical sophistication. It also reveals strategic technology bets they've made that influence how they'll evaluate your solution.

Traditional research methods for this layer involve reading company websites, press releases, analyst reports, and earnings calls. This takes hours and still gives you an incomplete picture.

AI-enhanced research can analyse thousands of data points in minutes:

- News articles and press mention to identify current challenges and initiatives
- Job postings to reveal what capabilities they're building (if they're hiring lots of data engineers, they're investing in analytics)
- Technology stack intelligence (tools like BuiltWith, Datanyze) to see what they're already using
- Financial data analysis to understand health, growth trajectory, and spending patterns
- Social media and review sites to gauge market perception and customer satisfaction

AI doesn't just gather this information. It synthesises it into a coherent picture of organisational context. Instead of reading 50 articles, you get a summary: "TechCorp is rapidly scaling internationally (8 new offices in 18 months), recently secured $50M Series C funding, hiring heavily in sales and marketing roles, and facing increasing competition from low-cost overseas competitors. Strategic priority appears to be operational scale while maintaining service quality."

That's actionable context you can build a conversation around.

Layer 2: Stakeholder Dynamics

At this level, we are understanding the people involved in the decision, how they relate to each other, and what organisational dynamics influence choices.

What you need to understand:

Decision-making structure: Who decides, who influences, who implements, and who can veto? In most B2B sales, the org chart is a polite fiction. Real decision authority doesn't follow titles. Understanding actual power dynamics—who defers to whom, who has the CEO's ear, who controls budget in practice—is essential for deal navigation.

Stakeholder alignment: Do the key players agree on priorities and approach, or are there competing agendas? A VP of Sales and a CFO might have very different views on whether to invest in new technology. The CTO might want advanced capabilities while the COO wants proven reliability. These tensions determine whether your deal moves forward smoothly or gets caught in internal politics.

Change capacity: How much organisational change can they absorb right now? A company in the middle of a major ERP implementation has limited capacity for additional change, regardless of how compelling your solution is. Understanding their change fatigue and competing initiatives helps you position appropriately.

Champions and blockers: Who will actively advocate for your solution, and who will resist? Champions aren't just people who like you. They're individuals with personal motivation to see your solution succeed and enough influence to drive it forward. Blockers aren't necessarily against you personally; they might have competing priorities, relationships with alternative vendors, or concerns about how your solution affects their power or responsibilities.

Influence networks: Who listens to whom? When the CTO recommends a vendor, does the CFO typically support that, or challenge it? When the VP of Operations raises concerns, does the CEO take them seriously? Understanding these influence patterns tells you how to build consensus.

Traditional stakeholder research involves conversations, org charts, and hopefully someone tells you who really matters. It's slow, incomplete, and biased toward who you happen to talk to first.

AI-enhanced stakeholder intelligence uses multiple data sources:

- **Network analysis:** Tools like LinkedIn Sales Navigator map who knows whom, how long they've worked together, and shared connections. AI can identify relationships that aren't obvious from titles.

- **Communication pattern analysis:** If you're using conversation intelligence tools, AI can analyse which stakeholders speak most in meetings, who asks challenging questions, who defers to others, and who makes definitive statements. This reveals actual influence regardless of title.

- **Historical deal analysis:** AI can identify patterns from previous deals: "In financial services companies this size, IT typically has veto power on security architecture regardless of who drives the initial evaluation."

- **Social media and public signals:** Blog posts, conference talks, social media activity, and industry involvement reveal individuals' priorities, expertise, and influence in their organisations and industries.

Instead of guessing at stakeholder dynamics, you can build evidence-based maps of who matters, how they relate, and what motivates them.

Layer 3: Individual Motivations

This is the micro level. Understanding each key stakeholder as an individual person with personal goals, career incentives, preferences, concerns, and cognitive patterns that influence how they make decisions.

What you need to understand:

Personal objectives: What is this individual trying to accomplish professionally? Are they trying to establish credibility in a new role? Protect a legacy they've built? Position themselves for promotion. Avoid risk that could damage their reputation? The same solution can appeal to or threaten different people based on their personal situations.

Career stage and trajectory: Someone early in their career might prioritise learning and proving themselves, making them open to innovative solutions. Someone late in career might prioritisestability and proven approaches. Someone clearly on a fast track might take risks for big wins. Someone plateaued might avoid anything that could go wrong.

Success metrics: How is this person evaluated? A sales leader measured on revenue growth has different priorities than one measured on margin or customer satisfaction. A CTO evaluated on system uptime thinks differently about new technology than one evaluated on innovation. Understanding their scorecard tells you what value propositions will connect.

Risk tolerance: Some individuals are naturally risk-averse; others are comfortable with uncertainty. Some have been burned by previous vendor relationships; others haven't. Risk tolerance isn't just personality. It's shaped by their situation, incentives, and experience.

Decision-making style: Does this person want detailed analysis and proof points, or do they trust intuition and move fast? Do they need to involve others and build consensus, or do they decide independently?

Do they focus on **strategic** vision or operational details? Matching your approach to their style dramatically improves effectiveness.

Communication preferences: How do they like to receive information? Dense documents? Executive summaries? Visual presentations? Conversational discussions? What medium works best. Email, phone calls, in-person meetings, video conferences?

Notice how this is where neuroscience and behavioural psychology become directly relevant to sales. Understanding how people make decisions—which we'll explore in the next section—allows you to align your approach with natural cognitive patterns rather than fighting against them.

Traditional individual understanding comes from conversations and maybe some background research. You learn by interacting with them over time. A slow, uncertain process.

AI-enhanced individual intelligence can analyse multiple signals:

- **Public content analysis:** Blog posts, articles, interviews, conference talks, and social media reveal priorities, thinking style, and concerns. AI can summarise: "This CTO writes frequently about balancing innovation with reliability, suggesting moderate risk tolerance and focus on proven-but-modern approaches."

- **Communication pattern analysis:** AI can analyse email exchanges, call transcripts, and meeting behaviours to identify decision-making patterns: "This CFO asks detailed questions about implementation costs in every interaction and returns to ROI calculations multiple times, suggesting need for rigorous financial justification."

- **Behavioural prediction:** Based on similar individuals' patterns, AI can suggest: "VPs of Marketing from consulting backgrounds typically prioritise differentiation and customer experience over operational efficiency."

The combination of these three layers—organisational context, stakeholder dynamics, and individual motivations—creates full understanding that transforms sales conversations. You're not just pitching a product; you're addressing real business challenges, steering through organisational complexity, and speaking to individual concerns in ways that hit home.

The Neuroscience of Decision-Making: How Buyers Actually Choose

Let's be honest: people don't make decisions the way they claim to. Ask any executive how they evaluate vendors, and they'll describe a rational process. Defining requirements, evaluating options, scoring vendors objectively, selecting the best fit. This is largely fiction.

Neuroscience research over the past two decades has revealed that human decision-making, even in professional contexts, is far less rational and far more emotional and pattern-based than we like to admit. Understanding these cognitive realities—and working with them rather than against them—is crucial for effective selling.

The Two-System Brain

Daniel Kahneman's work on "Thinking, Fast and Slow" revealed that we operate with two distinct cognitive systems:

System 1: Fast, automatic, emotional, unconscious. This is pattern recognition. Instantly recognising situations based on prior experience and making quick judgments. System 1 is always on, processing information in the background, generating gut feelings and intuitions.

System 2: Slow, deliberate, logical, conscious. This is analytical thinking. Carefully evaluating options, weighing pros and cons, calculating outcomes. System 2 requires effort and concentration; we can only use it for limited periods.

Here's what matters for sales:

System 1 *makes* most decisions. System 2 *rationalises* them afterward.

When a buyer says they need to "evaluate options carefully" and "build a business case," they're engaging System 2. But often, System 1 has already decided. They either feel confident in your solution or they don't. The evaluation process isn't genuinely open-ended; it's either finding rational justification for what System 1 already wants or finding rational reasons to reject what System 1 already doubts.

This isn't to say buyers are irrational. It's that rationality operates on top of emotional and intuitive foundations. Facts matter, but only once emotional and intuitive barriers are cleared.

Implications for selling:

Build emotional resonance first. Before diving into features and specifications, ensure the buyer feels confident in you, trusts your understanding of their situation, and believes you're genuinely trying to help. System 1 evaluates trustworthiness instantly. And if you fail that test, no amount of logical argument will overcome it.

Use stories and patterns, not just data. System 1 thinks in narratives and patterns, not spreadsheets. A case study from a similar company ("Company X faced exactly what you're facing and here's how it worked for them") lands more deeply than abstract ROI calculations. The pattern recognition system identifies similarity and generates confidence.

Make the safe choice feel obvious. System 1 is risk-averse by default. It looks for threats before opportunities. If your solution feels risky, uncertain, or complicated, System 1 will resist regardless of logical benefits. You need to make your solution feel like the safe, obvious path. Proven, low-risk, straightforward.

Provide rational justification for System 2. Once System 1 is on board, System 2 needs ammunition to justify the decision to others and to themselves. Provide the business case, ROI analysis, comparison charts, and implementation plans that let them rationalise their intuitive choice.

The Power of Social Proof and Authority

Another consistent finding from neuroscience and behavioural psychology: humans are deeply influenced by what others do and what authorities recommend. We're social creatures; our brains constantly monitor what our peers and respected figures are choosing.

This isn't weakness—it's an efficient cognitive shortcut. Evaluating every option from scratch is cognitively expensive. Following what similar others have successfully done, or what respected experts recommend, is usually a reliable strategy.

Implications for selling:

Use customer stories from similar companies. "Three other companies in your industry implemented our solution last year" carries more weight than any feature list. The more similar the reference customer is to your prospect, the more powerful the social proof.

Tap into authority and expertise. Industry analysts, respected consultants, and recognised experts influence decisions. If Gartner[4], Forrester, or respected industry figures validate your approach, prominently feature that. Your prospect's brain interprets this as "smart people who know more than I do think this is good, so it probably is."

Show momentum and adoption. Rapidly growing customer bases, industry awards, and visible adoption by recognizable brands all trigger social proof mechanisms. If everyone else is choosing you, your prospect's System 1 generates confidence that you're the safe choice.

---

[4] Gartner, "Hype Cycle for Revenue and Sales Technology, 2023," 2023.

Address competitive comparisons proactively. If competitors have social proof advantages (more customers, bigger brand), you can't ignore it. Address it directly by highlighting different criteria: "They're the biggest, which matters if you want the most generic solution. We're the best for companies like yours who need X specific capability."

Loss Aversion and Status Quo Bias

One of the most solid findings in behavioural economics: people feel losses roughly twice as strongly as equivalent gains. Losing $100 feels worse than gaining $100 feels good. This creates powerful status quo bias. Staying with current approaches feels safer than changing, even when change would create value.

Recognise that this is why selling new solutions is inherently harder than selling replacements or expansions of existing relationships. You're not just presenting benefits; you're asking people to accept risk, uncertainty, and potential loss.

Implications for selling:

Frame around what they're losing by not changing. Don't just talk about benefits your solution provides. Talk about costs of the status quo. "Every month you continue with your current approach, you're losing X in inefficiency" makes the status quo feel like a loss, not the safe baseline.

Reduce perceived risk of change. Pilots, phased rollouts, money-back guarantees, and success guarantees all reduce perceived downside risk. If changing feels low-risk, status quo bias weakens.

Make current pain vivid. Help prospects feel, not just intellectually acknowledge, their current problems. "Walk me through what happened last quarter when the forecast was off by 30%. How did that affect the business? How did you personally have to explain that?" Making current pain vivid shifts the emotional reference point, making change feel like relief rather than risk.

Position yourself as the lower-risk choice. If you're competing against alternatives (including status quo), don't position yourself as the "exciting" innovative option. Position yourself as the "safer" proven choice. Use customer references, proven track records, and risk mitigation to make your solution feel like the conservative decision.

The Anchoring Effect

When people make numeric estimates or evaluations, they're heavily influenced by whatever number they encounter first. If you ask someone "Do you think this project will cost more or less than $100,000?" their estimate will be very different than if you ask, "Do you think this project will cost more or less than $500,000?" The first number they hear becomes an anchor that influences all subsequent thinking.

Implications for selling:

Control the anchoring conversation. If you're selling a premium solution, don't start by talking about price. Start by establishing value, discussing the scale of the problem you're solving, or anchoring to the cost of the status quo. When you eventually discuss price, it's evaluated against those higher anchors rather than in isolation.

Use the right comparisons. If your solution costs $50K and you're competing against a $30K alternative, don't start with "$50K versus $30K." Start with "Your current inefficiency costs you $200K annually. Our solution reduces that by 60%, saving $120K per year, for a one-time investment of $50K." Now the anchor is $200K and $120K, not $50K versus $30K.

Break down large numbers differently. "$100K" feels very different from "$8,300 per month" or "$2,000 per week" or "$275 per day." Choose the framing that makes the investment feel proportionate to the value delivered.

The Paradox of Choice

More options don't always lead to better decisions or higher satisfaction. Research shows that excessive choice can lead to decision paralysis, anxiety, and lower satisfaction even when people do choose.

This is particularly relevant in complex B2B sales where buyers face dozens of vendors, hundreds of features, and countless implementation options.

Implications for selling:

Simplify the decision. Don't present every possible feature, package, and option. Present a clear recommendation based on their situation: "Given your priorities around X and Y, here's the configuration that makes sense for you."

Guide the decision process. Structure the evaluation into clear stages. First, we determine whether this type of solution is right for you. Then we configure the right approach. Then we address implementation. Breaking complex decisions into sequential, simpler choices reduces cognitive load and decision anxiety.

Limit comparisons. Don't encourage prospects to evaluate five vendors in parallel. Position yourself as different enough that comparison shopping doesn't make sense: "We're the solution for companies that prioritise X. If that's not your priority, we're not the right fit, and we can recommend alternatives."

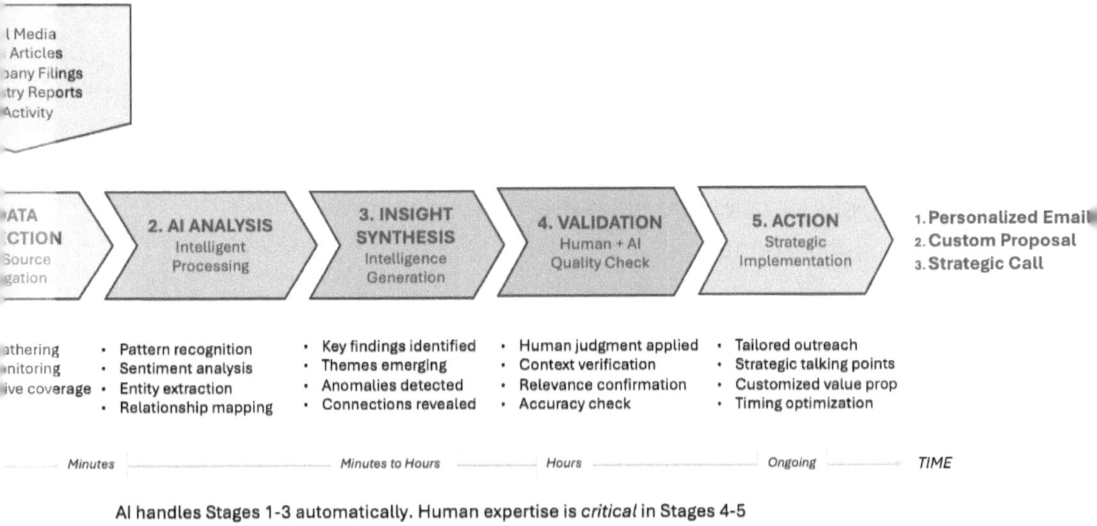

Figure 4.2: AI-Powered Research Workflow—
Five Stages from Raw Data to Strategic Intelligence

Understanding the three layers of customer intelligence and the cognitive patterns that drive decisions gives you the framework. AI gives you the capability to build that understanding at scale and speed that traditional research could never match.

Here's how to systematically apply AI for customer intelligence:

Step 1: Aggregate Data from Multiple Sources

Traditional research is limited by human capacity to gather and process information. You can read a company's website, some recent news, and maybe a few analyst reports. AI can ingest vastly more:

Public sources:

- Company websites, press releases, and investor relations materials
- News articles and press mention across thousands of publications
- Social media activity from company accounts and executives
- Industry analyst reports and market research
- Regulatory filings (e.g. 10-Ks, 10-Qs for public companies)
- Patent filings revealing R&D directions
- Job postings revealing capability buildouts

Proprietary databases:

- Technology stack intelligence (BuiltWith, Datanyze, HG Insights)
- Firmographic data (Dun & Bradstreet, ZoomInfo, Clearbit)
- Intent data showing what topics prospects are researching (Bombora, TechTarget)
- Financial data (PitchBook, Crunchbase for private companies)

Your own data:

- Previous interactions and relationship history
- Conversation intelligence from sales calls

- Email engagement patterns
- Website behaviour and content consumption

Relationship intelligence:

- LinkedIn connections showing who knows whom
- Board memberships and advisory relationships
- Shared investors or partners
- Conference attendance and speaking engagements

AI can process all of this in minutes. Work that would take a human researcher days or weeks.

Step 2: Synthesise Into Actionable Intelligence

But data aggregation isn't intelligence. The real value is synthesis. Turning thousands of disconnected facts into coherent understanding.

This is where modern AI excels. Natural language processing can read text across all these sources and identify:

Key themes and patterns: What topics come up repeatedly? What challenges are mentioned across multiple sources? What language does the company use to describe their priorities?

Significant changes: What's different now versus six months ago? New executives, new products, new partnerships, new challenges? Change creates buying windows. AI can spot them automatically.

Competitive context: Who are they compared to in media? Who do they partner with? Who are they hiring from? This reveals competitive dynamics and strategic positioning.

Sentiment and tone: Is the company confident and growing, or defensive and struggling? Sentiment analysis across sources reveals overall momentum and mindset.

Hidden connections: Relationships between people, shared backgrounds, common connections that aren't obvious from org charts alone.

Modern tools like Clay, Apollo, and specialized AI research assistants can automatically generate intelligence briefings that synthesise all this information into a readable summary. Instead of "here are 500 facts about this company," you get "here are the 5 key things you need to know to have a relevant conversation."

Step 3: Identify Behavioural Signals

Beyond static information, AI can track behavioural signals that indicate timing and intent:

Website behaviour: Which pages do prospects visit? How long do they spend? Do they look at pricing, case studies, or technical documentation? This reveals where they are in the buying journey and what concerns they have.

Content engagement: Which emails do they open? Which links do they click? Which assets do they download? Engagement patterns reveal interests and priorities.

Social media activity: What are prospects sharing, liking, or commenting on? This reveals what they're thinking about professionally.

Search behaviour: Intent data platforms track what topics companies are researching based on their employees' content consumption. If multiple people at a target account are researching "sales automation platforms," that's a strong buying signal.

Technology changes: When companies adopt new technologies, especially adjacent to your category, that often signals related needs. If they just implemented Salesforce, they might now need sales engagement tools that integrate with it.

AI can monitor these signals continuously and alert you when patterns indicate buying readiness: "TechCorp has had 15 people visit your website in the past week, with 6 looking at pricing and case studies. Three executives follow you on LinkedIn. Intent data shows increased research on your category. This suggests active evaluation."

Step 4: Predict and Recommend

The most advanced application of AI in customer understanding is predictive. Using patterns from historical data to suggest what's likely true or what's likely to happen, even when direct evidence isn't available.

For example:

> *"Based on similar companies in this industry, there's an 78% probability that IT will need to approve security architecture before purchase can proceed. Recommend engaging their CISO early."*

> *"Sales leaders from consulting backgrounds typically prioritise differentiation over efficiency. Recommend emphasising your unique methodology rather than operational benefits."*

> *"This company's tech stack suggests they value best-of-breed solutions over all-in-one suites. Position as specialized excellence rather than thorough platform."*

Notice how thise predictions aren't certain. But they're based on patterns across thousands of data points and give you high-probability starting assumptions rather than going in blind.

Turning Understanding into Conversations

Intelligence is only valuable if it changes how you engage with prospects. Here's how deep understanding transforms typical sales interactions:

Generic sales conversation:

Rep: "Thanks for taking the time. I wanted to learn about your business and see if our solution might be a fit. Can you tell me about your current challenges around [product category]?"

Prospect: *Provides generic overview, testing whether this rep is worth their time*

Rep: "Interesting. And what would solving that look like for you?"

Prospect: *Gives expected answer, still not particularly engaged*

This is fine—but it's transactional and commodity-level engagement.

Intelligence-informed conversation:

Rep: "I know you just launched your new product line in Europe last quarter, and based on your job postings, it looks like you're scaling the sales team there pretty aggressively. I imagine that's creating pressure on your sales operation's infrastructure. Most companies we work with face bandwidth issues when they're scaling into new markets that fast while maintaining growth in existing regions. Is that consistent with what you're experiencing?"

Prospect: *Immediately engaged because you clearly understand their situation* "Actually yes. We're finding that our processes that worked fine at 50 reps completely break down at 150 reps across three regions..."

Notably, the prospect immediately raises the conversation because you've demonstrated you understand their context. Now you're having a strategic discussion, not an interrogation.

Personalised stakeholder engagement:

Instead of the same pitch to every stakeholder, you adjust based on individual understanding:

To the VP of Sales (focused on growth, evaluated on revenue): "Your team's grown 60% in the last year based on your LinkedIn updates. The challenge I typically see at that growth rate is maintaining consistency.

New reps ramping slower than planned, top performers' best practices not scaling across the team. Our clients in similar growth situations typically see 30-40% faster ramp time because we systematise what your best reps do naturally."

To the CFO (focused on efficiency, evaluated on margins): "I saw your last earnings call mentioned pressure on sales and marketing efficiency as you scale. That connects with what we're seeing broadly in SaaS companies at your stage. The companies we work with typically find that sales automation delivers 3:1 ROI in the first year, primarily through rep productivity gains rather than headcount reduction. I'd be interested in walking through how that math might work in your context."

To the CTO (focused on architecture, evaluated on reliability and integration): "I saw you've recently standardised on Salesforce and invested pretty heavily in your data infrastructure based on your engineering blog posts. That suggests you're thinking about your tech stack architecturally, not just tactically. Our platform was designed from the ground up to integrate with Salesforce natively. We use their platform APIs rather than just syncing data, which means reliability and performance are much better than bolt-on tools."

Same solution but positioned completely differently based on what you understand about each person's priorities and evaluation criteria.

The Understanding Feedback Loop

The most important aspect of customer understanding: it's not a one-time research project. It's a continuous process of hypothesis-testing and refinement.

You form initial understanding from research. Then every conversation provides feedback. Do they respond as predicted? Are they concerned about what you anticipated? Do they make decisions the way you expected? This feedback refines your understanding.

Document what you learn:

- Capture insights from every customer interaction
- Note what connects and what doesn't
- Record unexpected concerns or priorities
- Update your understanding as you learn more

Feed insights back to AI:

- When you discover something conversation intelligence didn't catch, add it to the record
- When predictions are wrong, understand why so models improve
- When you win or lose deals, document what understanding was accurate or inaccurate

Over time, your understanding becomes increasingly sophisticated. Not just for individual accounts, but across patterns: "CFOs in financial services always care about X" or "Companies expanding internationally face Y challenge that they don't initially mention."

This compounding intelligence is where elite sellers separate from average ones. Average sellers start every conversation from scratch. Elite sellers start with deep, pattern-informed understanding that immediately establishes credibility and accelerates trust.

Common Understanding Mistakes

Before we conclude, let's address the most common ways understanding goes wrong:

Mistake 1: Confusing Data with Understanding

Having lots of information doesn't mean you understand anything. I've watched reps show up to calls with 20-page research dossiers and still fail to have relevant conversations because they hadn't synthesised the data into insight.

Understanding means answering: "Why does this matter? What does this mean? How does this affect their decisions?" Data is input; understanding is output.

Mistake 2: Assuming Rather Than Validating

AI can predict with high probability, but it's not certain. "Based on patterns, this CFO probably cares about X" is a hypothesis, not a fact. Use understanding to inform initial approach but validate through actual conversation. If your hypothesis is wrong, update it. Don't cling to what research suggested over what the prospect tells you.

Mistake 3: Overwhelming Prospects with Your Knowledge

Just because you've learned a lot about their business doesn't mean you should demonstrate it all at once. Nobody likes feeling studied. Use understanding to ask better questions and make relevant points, but don't make prospects feel like you've been stalking them.

There's a fine line between "you clearly understand our business" and "this is creepy. How do you know all this?"

Mistake 4: Understanding Without Action

Understanding that doesn't change your behaviour is wasted effort. If you learn that a prospect prioritises X, but you still pitch Y, your research didn't matter. The point isn't to be knowledgeable. It's to be effective. Use understanding to adjust your approach.

Putting Understanding into Practice

Understanding is the foundation that makes everything else in QUANTUM work effectively. You can't anticipate objections without understanding motivations. You can't steer through stakeholder dynamics without understanding relationships. You can't tailor effectively without understanding preferences. You can't measure accurately without understanding what drives outcomes.

In the next chapter, we'll explore Anticipate—how to use the deep understanding you've developed to predict objections, identify risks, and prepare responses before you ever walk into critical conversations. Because understanding what is doesn't help unless you can also predict what's coming.

CHAPTER OUTCOME

☐ A repeatable "customer intelligence brief" template (one page)

☐ A stakeholder map for one real target account (roles, motivations, likely objections)

☐ A list of 5–10 intelligence inputs you will collect every time (signals, news, financials, priorities, tech stack, etc.)

☐ A method for turning information into actionable insight (so you don't just research)

☐ A set of discovery prompts that move from generic questions to insight-led questions

☐ A list of common "understanding errors" you personally fall into (and how you'll prevent them)

☐ One "conversation hypothesis" for your next key account (what you believe is true + what you must validate)

CHAPTER 5

Anticipate – Predictive Sales Intelligence

By the end of this chapter, you will be able to:

- Explain the commercial cost of reactive selling and what anticipation changes.
- Work across the three domains of anticipation: objections, risks, and opportunities.
- Build and operationalise an objection library and map likely objections to prospect characteristics.
- Create a risk assessment matrix, monitor leading indicators, and run scenario planning ("what-if engine").
- Close the loop with post-deal analysis and AI feedback, avoiding "prediction without action."

How decisions are really made . And why most sales effort misses the point

I once watched a seven-figure deal collapse in the final hour because the CFO raised an objection about integration costs that blindsided the entire sales team. The irony? Three other companies in the same industry had raised identical concerns in previous quarters. The pattern was obvious. To anyone who'd been paying attention. But the team was so focused on crafting the perfect proposal and building relationships with the VP of Sales that they never anticipated the financial executive would torpedo the deal over implementation risk.

On closer inspection, this happens more often than anyone wants to admit. Salespeople work deals for months, invest countless hours building consensus, address every stated concern, and then get hit with an objection they never saw coming. The deal either dies on the spot or enters an endless loop of "we need more information" while the prospect goes cold.

Here's what separates average sellers from elite ones: elite sellers rarely get surprised. They don't have better luck or easier prospects. They've simply learned to anticipate what's coming before it arrives. And in the AI age, this capability isn't just about experience and intuition anymore. It's about combining human judgment with machine intelligence to predict objections, identify risks, and prepare responses before you ever walk into the room.

This is the Anticipate stage of the QUANTUM Framework. If Qualify gets you focused on the right prospects and Understand gives you deep intelligence about them, Anticipate is what keeps you three steps ahead throughout the sales cycle.

The Cost of Being Reactive

Most sales professionals operate in reactive mode. A prospect raises a concern, and you address it. A competitor makes a move, and you respond. A stakeholder surfaces a new requirement, and you scramble to accommodate it. This feels like normal sales work because it's what everyone does.

But reactive selling has a massive hidden cost: time. Every time you're surprised by an objection, you lose momentum. The deal stalls while you regroup, gather information, and formulate a response. Even if you ultimately address the concern successfully, you've added days or weeks to the sales cycle. Multiply this across dozens of deals, and reactive selling is costing you significant revenue. Not because you're losing deals, but because you're closing them slower than necessary.

There's also a credibility cost. When prospects raise concerns, you haven't anticipated, it signals you don't understand their business as well as you should. They start to wonder: if you missed this obvious issue, what else are you missing? Trust erodes, even if you handle the objection well.

The companies that consistently outperform their targets don't just react better. They anticipate more effectively. They've learned to spot patterns, predict objections, and identify risks before they become problems. And increasingly, they're using AI to do this at a level that human analysis alone simply can't match.

How AI Changes Anticipation

Traditional objection handling relies on experience. You work enough deals; you start to recognise patterns. You hear the same concerns repeatedly. You develop a mental library of objections and responses. This works. To a point.

Across industries, the limitation is scale. A human can remember maybe 50-100 deals in detail. An experienced seller might recognise a dozen common objection patterns. But AI can analyse thousands of deals simultaneously, identify subtle patterns humans would never catch, and surface correlations that aren't obvious from individual transactions.

Here's a real example: Salesforce analysed its win/loss data and discovered that deals where the champion changed roles mid-cycle had a 73% higher chance of stalling or being lost. But only if the change happened in certain stages of the buying process. Early changes increased win rates slightly, presumably because the new person wanted to make their mark with a significant decision. Late-stage changes were devastating.

No individual salesperson would ever notice this pattern across enough deals to make it actionable. But AI spotted it immediately, which allowed Salesforce to build early warning systems that flagged deals when champions changed roles late in the cycle. Sales teams could then proactively address the transition risk before it killed momentum.

This is the power of AI-enabled anticipation: finding patterns in massive datasets that reveal what's likely to happen next, giving you time to prepare or intervene before issues become crises.

The Three Domains of Anticipation

Effective anticipation operates across three distinct domains: objections, risks, and opportunities. Most salespeople focus almost exclusively on the first and ignore the other two. Elite sellers—and AI-enhanced sales operations—monitor all three simultaneously.

Domain 1: Objection Anticipation

Objections are the most obvious thing to anticipate because they directly block deals. But most salespeople approach objection handling backwards. They wait for the objection to be voiced, then try to overcome it. This puts you in a defensive position, explaining why their concern isn't valid or isn't as significant as they think.

Far more effective: anticipate the objection and address it before it's raised. This completely changes the dynamic. Instead of defending your position, you're demonstrating that you understand their business well enough to know what they'll be concerned about.

When I was selling enterprise software to financial services firms, we knew procurement teams would object to our security architecture. Every single time. It was the most predictable objection in our pipeline.

Early in my career, I'd wait for them to raise it, then scramble to get our security team on a call to address their concerns. This added two weeks to every deal while we scheduled meetings and answered technical questions.

Then we started anticipating it. In our initial proposals, we included a dedicated security section that proactively addressed the five most common concerns before they were raised: data encryption standards, compliance certifications, penetration testing results, disaster recovery protocols, and third-party security audits. We also offered to have our CISO meet with their security team early in the process, before they even asked.

The impact was dramatic. Not only did deals move faster, but we had fewer security objections overall. By addressing concerns proactively, we established credibility that made prospects trust our platform was more secure than alternatives, not less.

AI amplifies this approach by analysing which objections arise in which situations. Instead of guessing based on general experience, you get specific predictions: "Based on this prospect's industry, company size, and current tech stack, there's an 82% probability they'll object to implementation timeline and a 67% probability they'll challenge ongoing support costs."

Armed with this intelligence, you can proactively address these specific concerns in your proposal, demonstration, and business case before they ever become objections.

Domain 2: Risk Anticipation

While objections are concerns prospects voice, risks are problems that might derail deals whether anyone talks about them or not. This includes things like:

- Budget getting reallocated mid-cycle
- Key stakeholders leaving the company
- Competitive threats you don't know about yet
- Technical requirements that emerge late in the process

- Legal or compliance issues that block procurement
- Internal political dynamics that kill consensus

In effect, thise risks exist whether you see them coming or not. The question is whether you identify and address them early or get blindsided when they torpedo your deal.

Traditional risk management in sales relies on asking questions and hoping prospects tell you the truth. "Is the budget confirmed?" "Who else is involved in this decision?" "Are there any technical requirements we should know about?"

The problem: prospects often don't know the answers themselves. The marketing VP genuinely believes she has budget authority, not realising the CFO will scrutinize any purchase over $100K. The IT director says he's the decision-maker, unaware that the CTO has veto power on new vendors. The champion assures you there are no other solutions being evaluated, not knowing that a different business unit is already piloting your competitor.

AI-powered risk anticipation works by identifying patterns that predict problems, even when prospects aren't aware of them. For example:

- **Budget risk:** AI can flag that similar deals in this industry/company size typically get escalated to finance for approval, even when initial contacts claim they have authority. You can then proactively build financial justification into your business case rather than being surprised when CFO approval is suddenly required.
- **Stakeholder risk:** AI can map typical decision-making structures and flag when you haven't engaged with roles that historically influence the decision. If you're selling HR software and haven't talked to IT yet, AI might predict a 68% chance of implementation concerns being raised late in the process.
- **Competitive risk:** AI can monitor job postings, news about your prospect, social media mentions, and tech stack changes to

identify signs a competitor might already be engaged, even if your champion hasn't mentioned it.

- **Timeline risk:** By analysing historical patterns, AI can predict that deals of this size and complexity typically take 4-6 months to close, alerting you when a prospect's aggressive timeline (they want to close in 30 days) is likely unrealistic. This lets you manage expectations early rather than watching the deal slip repeatedly.

key insight: risks often follow predictable patterns across similar situations. AI can spot these patterns and alert you to likely problems based on characteristics of your deal, even when the specific risk hasn't materialised yet.

Domain 3: Opportunity Anticipation

This is the domain most salespeople completely ignore, yet it's potentially the most valuable. While you're focused on objections and risks—defensive plays—AI can also identify opportunities you're missing ways to expand the deal, stakeholders you should engage, adjacent needs you could address, or timing windows you should exploit.

For example, Gong.io's AI analyses sales calls and can identify when prospects mention adjacent problems your company could solve, even if the salesperson doesn't catch it in the moment. A prospect might mention in passing that they're struggling with customer onboarding, and the AI flags this as an expansion opportunity for your onboarding solution. Something the rep might have missed while focused on closing the initial deal.

Similarly, AI can identify timing opportunities by monitoring company announcements, leadership changes, funding rounds, or market events that create buying windows. When a company announces a new product line, that might signal they'll need new sales tools to support it. When a new CMO joins, they typically want to make their mark with new technology decisions. AI can surface these signals and alert you to strike while timing is optimal.

LinkedIn's Sales Navigator uses AI to surface "trigger events". Company changes that create sales opportunities. When a target account gets acquired, goes through a merger, or opens a new office, these events create needs for new solutions. Rather than waiting to stumble across these opportunities, AI proactively identifies them so you can reach out at exactly the right moment.

The Objection Anticipation Framework

Let's get tactical. Here's a systematic framework for anticipating objections before they arise:

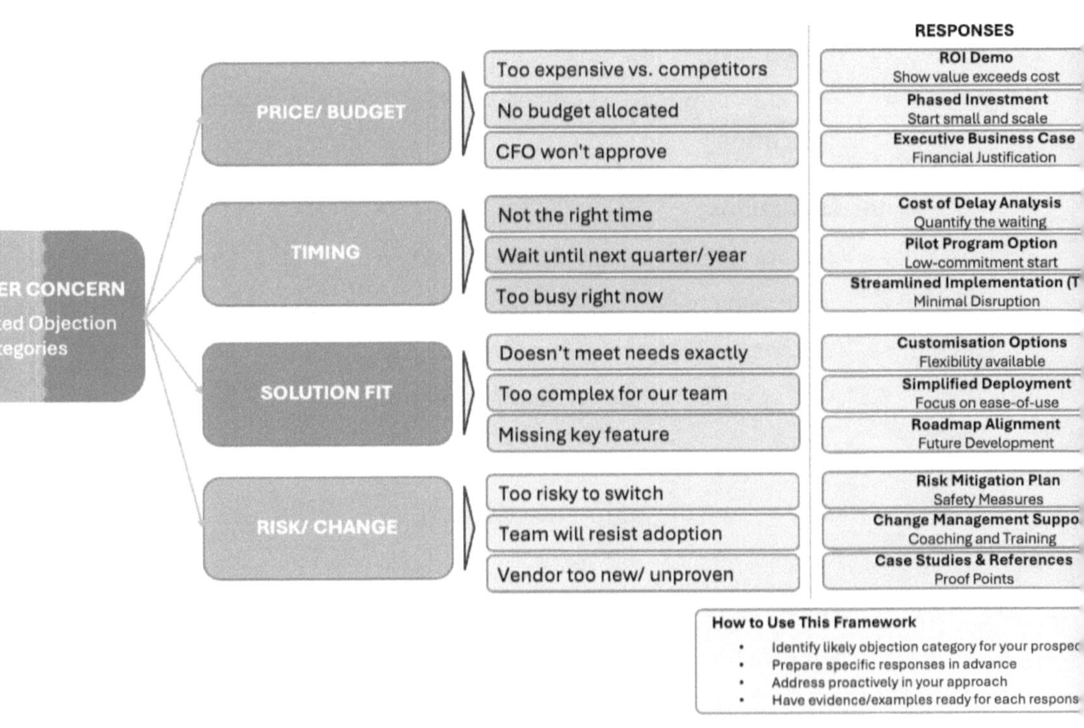

Figure 5.1: Objection Anticipation Decision Tree with Response Strategies - Systematic Preparation for Common Sales Objections

Step 1: Build Your Objection Library

Start by cataloguing every objection you've encountered in the last 50 deals. Don't just list them. Categorise them:

Price/Budget Objections:

- "Too expensive compared to alternatives"
- "No budget allocated for this year"
- "ROI timeline is too long"
- "CFO won't approve purchases over $X without additional justification"

Timing Objections:

- "Not the right time, revisit next quarter"
- "Too much change happening, can't take on something new"
- "Current contract doesn't expire for 6 months"
- "Busy season coming up, can't implement now"

Solution Fit Objections:

- "Missing specific feature we need"
- "Too complex for our team"
- "Doesn't integrate with our existing tech stack"
- "Not designed for our industry/use case"

Risk/Change Objections:

- "Too risky to switch from current solution"
- "Team will resist new tool"
- "Implementation seems too disruptive"
- "Vendor is too new/unproven"

For each objection, note:

- Which stage of the sales cycle it typically appears
- Which persona (role) usually raises it
- What ultimately resolved it (or didn't)

- How much time it added to the deal cycle

This library becomes the foundation for anticipation. You're systematizing the experience that lives in your head so both you and AI can pattern-match against it.

Step 2: Feed Your Library to AI

Modern CRM systems and conversation intelligence tools can ingest this objection library and then automatically flag when similar patterns emerge in new deals. Tools like Gong, Chorus.ai, and Clari can analyse your sales calls and identify when prospects are hinting at objections even before explicitly stating them.

For example, when a prospect says, "We're really focused on cost control this quarter," AI can flag this as an early signal of a price objection coming. When someone mentions "We've been burned by implementations before," that's a predictive signal of risk/change objections. The AI doesn't wait for the explicit objection. It identifies the early warning signs.

You can then proactively address these emerging concerns before they crystallise into full objections that stall your deal.

Step 3: Map Objections to Prospect Characteristics

Not all objections are equally likely for all prospects. The objections a Fortune 500 enterprise raises are different from those of a 50-person startup. Financial services companies object to different things than retail companies.

AI can analyse which objections correlate with which prospect characteristics:

- Industry
- Company size
- Current tech stack
- Stage of company growth

- Geographic region
- Regulatory environment

When you're working a new deal, AI can predict: "Based on this prospect being a 500-person healthcare company in the Northeast using Salesforce, there's a 78% chance they'll object to HIPAA compliance complexity and a 65% chance they'll push back on integration effort."

Armed with these predictions, you can build your proposal, demo, and business case to proactively address these specific concerns rather than waiting for them to be raised.

Step 4: Build Pre-emptive Response Strategies

For each major objection category, develop response strategies you can deploy before the objection is even raised:

For price objections: Build ROI calculators, comparison tools, and financing options into your standard proposal. Don't wait for them to say it's expensive. Show upfront why the investment makes financial sense.

For timing objections: Address implementation impact, change management, and resource requirements early. Show you understand their concerns about disruption and have plans to minimise it.

For solution fit objections: Proactively discuss the most requested features, explain your product roadmap, and address integration considerations in your demo.

For risk/change objections: Include customer success stories from similar companies, highlight your implementation support, and offer pilot programs or phased rollouts to reduce perceived risk.

The goal isn't to eliminate objections. Prospects will always have concerns. The goal is to shift from reactive (waiting for objections, then responding) to proactive (addressing likely concerns before they're raised, demonstrating you understand their business).

Risk Pattern Recognition

Beyond specific objections, successful deals require identifying and mitigating broader risks. Here's how to systematically anticipate risk:

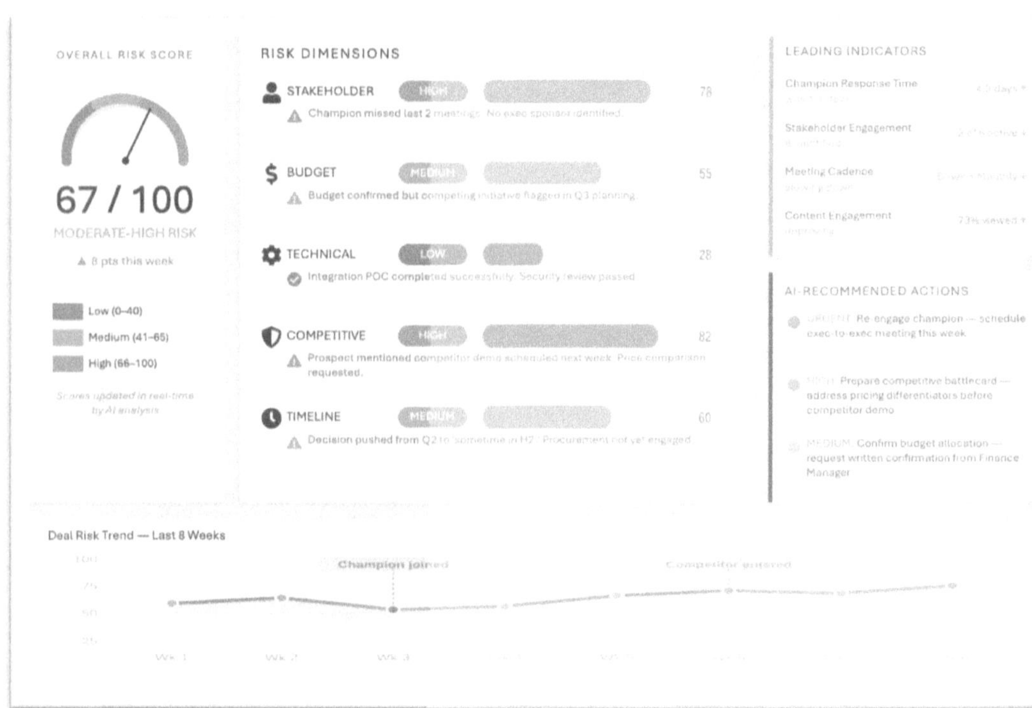

Figure 5.2: AI-Powered Deal Risk Assessment Dashboard - $420k Enterprise Platform De...

Create a Risk Assessment Matrix

For every active deal, evaluate these risk dimensions:

Stakeholder Risk:

- Have you identified all key decision-makers and influencers?
- Is your champion strong enough to drive consensus?
- Are there hidden blockers you haven't engaged?
- What happens if your champion leaves or changes roles?

Budget Risk:

- Is budget truly allocated, or just "available"?
- Who has final spending authority?
- Are there competing priorities for the same funds?
- Could budget get reallocated if priorities shift?

Technical Risk:

- Are there integration challenges you haven't fully scoped?
- Could technical requirements emerge that your solution doesn't meet?
- Is the implementation timeline realistic given their resources?
- Are there security/compliance requirements that could block procurement?

Competitive Risk:

- Are other solutions being actively evaluated?
- Could the prospect decide to build internally rather than buy?
- Is there an incumbent vendor with a strong relationship?
- Could a competitor undercut on price or offer better terms?

Timeline Risk:

- Is the prospect's timeline realistic based on their decision-making process?

- Are there external events (fiscal year end, compliance deadlines) creating urgency?
- Could the deal slip if key stakeholders are unavailable?
- Are there dependencies (other initiatives, budget approvals) that could delay decisions?

For each dimension, assign a risk score: Low, Medium, or High. Then use AI to identify patterns: which combinations of risk factors historically predict deal delays or losses?

For example, you might discover that deals with High Stakeholder Risk + High Timeline Risk have a 68% probability of slipping by at least one quarter. This lets you proactively slow down overly aggressive timelines and invest in broader stakeholder engagement before the deal stalls.

Monitor Leading Indicators

AI can track behavioural signals that predict risk is increasing:

- **Engagement drop-off:** When champion responsiveness declines, deal risk increases. AI can flag when email response times lengthen or meeting acceptance rates drop.
- **Stakeholder churn:** When key contacts leave calls early, skip meetings, or hand you off to more junior people, these signal declining priority. AI can track meeting attendance patterns and alert you to these shifts.
- **Competitor mentions:** When prospects start asking about feature comparisons or bringing up competitor names, competitive risk is rising. Conversation intelligence tools can flag these mentions in real-time.
- **Legal/procurement delays:** When deals enter legal review or procurement, velocity often predicts outcome. Deals that move quickly through these stages close; deals that slow down often die. AI can identify when these processes are taking longer than historical norms, signalling elevated risk.

The key is catching these signals early, when you still have time to intervene, rather than discovering problems only when deals fail to close as expected.

Scenario Planning: Preparing for Multiple Futures

Elite sellers don't just predict the most likely outcome. They prepare for multiple scenarios. Before important meetings or milestones, they mentally rehearse different directions the conversation could go and prepare responses for each.

AI makes this scenario planning more systematic and complete. Here's how:

The "What If" Engine

Before significant customer interactions—executive presentations, final proposals, negotiation meetings—use AI to generate likely scenarios:

Scenario 1: The Sceptical CFO

- Prediction: CFO questions ROI timeline and asks for more conservative projections
- Preparation: Have detailed financial models with sensitivity analysis ready. Prepare case studies from similar companies showing actual results. Be ready to discuss phased implementation with milestone-based payments.

Scenario 2: The Competitive Challenge

- Prediction: Prospect mentions competitor X is offering better pricing or specific feature advantages
- Preparation: Research competitor's actual pricing and limitations. Prepare comparison showing your total cost of ownership is lower despite higher upfront cost. Have customer references who switched from that competitor ready to share.

Scenario 3: The Implementation Concern

- Prediction: IT team raises concerns about integration complexity and resource requirements
- Preparation: Have detailed implementation plan with specific resource estimates. Propose implementation partners or managed services options. Show examples of similar technical environments you've integrated with successfully.

Scenario 4: The Timing Objection

- Prediction: Stakeholders want to delay decision until next budget cycle or current contract expires
- Preparation: Quantify cost of delay—revenue lost or inefficiency continuing during waiting period. Propose pilot program that starts small and expands. Offer pricing incentive for committing now with delayed implementation.

By preparing for multiple scenarios, you're never caught flat-footed. When the CFO does raise ROI concerns, you don't scramble to respond. You smoothly address it because you anticipated the possibility and prepared accordingly.

Using AI for Scenario Generation

Conversation intelligence platforms can analyse your past calls and identify common conversational patterns:

- What objections typically follow which statements?
- What questions do prospects ask after seeing specific demos?
- What concerns arise when stakeholders join calls?

Practically speaking, this historical pattern analysis lets AI suggest likely scenarios before your next customer interaction. Instead of guessing what might happen, you're working from data about what typically happens in similar situations.

The Anticipation Feedback Loop

At its simplest, the real power of anticipation comes from continuous learning. Every deal teaches you something about what to expect next time. The question is whether you're systematically capturing and applying those lessons.

Here's how to build an anticipation feedback loop:

Post-Deal Analysis

After every won or lost deal, document:

- Which objections arose vs. what you anticipated
- Which risks materialised vs. what you predicted
- Which opportunities you identified vs. what you missed
- How accurate your scenario planning was

This isn't about judging yourself. It's about refining your predictive models. If you anticipated a price objection that never came, what was different about this deal? If a risk you didn't foresee killed the deal, what signals could have predicted it?

Feed Insights Back to AI

Modern CRM systems can track this feedback and use it to improve predictions. When you document that a certain type of prospect raised specific objections, the AI's models get more accurate for future similar prospects.

Over time, your anticipation capability compounds. Six months into systematic anticipation practice, you're not just reacting faster. You're encountering fewer surprises because both you and your AI tools have learned to spot early warning signs and emerging patterns.

Common Anticipation Mistakes

Before we move on, let's address the most common ways anticipation goes wrong:

Mistake 1: Over-Preparing for Low-Probability Scenarios

Just because something could happen doesn't mean it's worth extensive preparation. If AI predicts a 12% chance of a specific objection, don't spend hours preparing for it. Focus your energy on high-probability scenarios and have lightweight responses ready for edge cases.

Mistake 2: Addressing Objections That Haven't Arisen

There's a balance here. Proactively addressing likely concerns demonstrates understanding. But raising objections that prospects weren't even thinking about can create doubt. Use judgment: address obvious concerns proactively, but don't invent problems.

Mistake 3: Trusting AI Predictions Without Context

AI can tell you that 78% of similar prospects raised a specific objection. But if you've already addressed the underlying issue through your unique approach, that historical pattern may not apply. Use AI predictions as input, not commandments.

Mistake 4: Anticipating But Not Acting

Knowing an objection is coming doesn't help if you don't do anything about it. Anticipation is only valuable when it changes your behaviour. You adjust your proposal, proactively address concerns, or prepare specific responses.

Putting Anticipation into Practice

Let's bring this together with a practical example of how anticipation works in a real deal cycle.

You're selling marketing automation software to a mid-sized B2B company. Here's how anticipation operates at each stage:

Early Stage (Week 1-2):

- AI analysis of similar prospects predicts 82% will object to implementation complexity and 71% will question ROI timeline
- You proactively build simple implementation plan and conservative ROI model into initial proposal
- During demo, you specifically show ease of setup and quick time-to-value rather than just feature depth

Mid Stage (Week 3-5):

- Conversation intelligence flags that VP of Marketing mentioned "getting burned by previous software purchases"
- You recognise this as early signal of risk/change objection
- Proactively offer pilot program with 3 initial users before full rollout, reducing perceived risk
- Share customer success story from similar company that had difficult prior experience but succeeded with your solution

Late Stage (Week 6-8):

- AI flags that champion hasn't engaged CFO yet, which historically predicts budget approval delays
- You proactively suggest involving finance team, offering to present business case directly
- During CFO meeting, you anticipate questions about total cost of ownership vs. just licence fees
- You arrive prepared with 3-year TCO analysis including implementation, training, and ongoing costs

Final Stage (Week 9-10):

- Deal unexpectedly slows down; champion becomes less responsive

- AI flags similar engagement drops patterns that historically predict competitive threat
- You directly but tactfully ask champion if other solutions are being evaluated
- Champion admits competitor approached them with aggressive pricing
- Because you anticipated this possibility, you've already prepared value justification that goes beyond price, emphasising your superior support and integration capabilities

Result: Deal closes in 10 weeks instead of slipping to 16+ weeks because you proactively addressed objections and risks before they became blocking issues.

The Anticipation Advantage

Here's what changes when you systematically anticipate rather than react:

- **Your deals move faster** because objections get addressed before they stall momentum.
- **Your win rates improve** because risks get mitigated before they kill deals.
- **Your credibility increases** because prospects see you understand their business well enough to predict their concerns.
- **Your forecast accuracy improves** because you can identify early warning signs that deals are at risk.
- **Your stress decreases** because you're rarely caught by surprise. You've already thought through likely scenarios and prepared responses.

Most crucially, you shift from constantly playing defence to playing offence. Instead of spending your time reacting to problems, you're steering deals proactively toward successful outcomes.

In the next chapter, we'll look at Navigate—how to orchestrate complex deals through organisational politics and decision-making mazes once you understand what's coming and have prepared accordingly. Anticipation tells you what's likely to happen; Navigation is about making sure the right things happen when they matter most.

<table>
<tr><td>

CHAPTER OUTCOME

☐ A working Objection Library (top objections + best evidence-based responses)

☐ A simple risk matrix (deal risks you will monitor + leading indicators)

☐ A "pre-mortem" template you can run early in deals ("why might this fail?")

☐ A scenario planning checklist (what-if branches for pricing, procurement, legal, competition, timing)

☐ A method for capturing learnings post-deal (wins and losses)

☐ A plan for feeding learnings back into your library (so it improves over time)

☐ One "anticipation routine" you'll run weekly (20–30 minutes)

</td></tr>
</table>

CHAPTER 6

Navigate – Strategic Deal Orchestration

Steering complex sales through organisational politics and decision-making mazes

> **By the end of this chapter, you will be able to:**
>
> - Map stakeholder networks beyond the org chart and identify hidden decision-makers and influence patterns.
> - Execute multi-threading properly (champion strategy, coalition building, and blocker management).
> - Run power mapping to locate true economic authority and decision control.
> - Drive momentum by managing both formal and informal decision processes stage-by-stage.
> - Use AI-powered navigation intelligence (relationship mapping, org intelligence, deal risk prediction, competitive signals).

I once spent four months working what seemed like a straightforward deal. The VP of Sales loved our solution. We'd done multiple demos. The business case was rock solid. The timeline was clear. By every traditional measure, this deal should have closed.

Then, two weeks before the expected signature, everything went silent. No responses to emails. Calls went unreturned. Meetings were cancelled. The deal just... vanished.

Three months later, I found out what happened. The CTO—someone I'd never even spoken to—had vetoed the purchase. Apparently, he had a strategic relationship with a competing vendor and wasn't willing to bring in an alternative solution, regardless of what the VP of Sales wanted. The VP didn't have the authority or political capital to override him, so rather than having an awkward conversation with me about internal politics, he just ghosted.

The deal was dead before I even knew there was a problem. And the worst part? There were warning signs I'd completely missed. I was single threaded with one champion in one department. I never mapped the broader organisational dynamics. I never understood the decision-making structure. I never identified who could kill the deal, only who could advocate for it.

I'd qualified well. I'd understood the explicit business need. But I completely failed to steer through the organisational complexity that determines whether deals close.

This is what the Work through stage of the QUANTUM Framework addresses. In simple sales—transactional purchases with single decision-makers—navigation doesn't matter much. But in complex B2B sales involving multiple stakeholders, competing priorities, and organisational politics, navigation capability separates deals that close from those that stall indefinitely.

You can have the best solution, the strongest business case, and enthusiastic champions, but if you can't steer through the organisational maze from initial interest to final signature, none of it matters.

Why Complex Deals Require Navigation

Let's start by acknowledging what makes B2B sales inherently complex:

Multiple stakeholders with different priorities: The VP of Sales cares about team productivity. The CFO cares about ROI and cash flow. The CTO cares about architecture and integration. The Legal team cares about risk and liability. The Procurement team cares about vendor management and contract terms. Your solution needs to work for all of them, but "work for" means different things to each.

Competing interests and agendas: Sometimes these stakeholders have genuinely conflicting goals. Sales wants flexibility and features. IT wants standardisation and security. Finance wants low cost. Marketing wants sophisticated capabilities. Your solution can't perfectly satisfy everyone, which means someone's needs get prioritised and someone's get deprioritised. A political decision, not a purely logical one.

Hidden power dynamics: The org chart shows formal reporting structure, but real influence doesn't flow through official channels. The person with the EVP title might defer to the Senior Director who has the CEO's ear. The person who seems junior might be a rising star whose opinion carries disproportionate weight. Understanding actual power dynamics requires investigation that goes far beyond LinkedIn titles.

Risk aversion and status quo bias: Organisations resist change by default. Even when your solution is objectively better, change creates risk. Implementation risk, adoption risk, reputation risk for whoever champions it. The default answer to "should we buy this?" is always "not yet" unless someone actively drives it forward.

Long, unpredictable decision cycles: Complex purchases involve legal review, security assessment, procurement negotiation, budget approval, technical validation, and multiple layers of management sign-off. Each stage can stall or reverse momentum. What looks like a 90-day deal can easily turn into 12 months.

Traditional sales training largely ignores this complexity. It teaches you to "find the decision-maker" and "build consensus" without acknowledging that in enterprise sales there often isn't a single decision-maker, and consensus isn't naturally achieved. It must be orchestrated through deliberate strategy.

In practice, this is where elite sellers distinguish themselves. They don't just pitch value. They actively guide organisational dynamics to move deals through complexity toward close.

Stakeholder Mapping: Understanding the Cast of Characters

Navigation starts with understanding who's involved and how they relate to each other. This is stakeholder mapping. Creating a thorough picture of everyone who influences the decision, their relationships to each other, and their individual priorities.

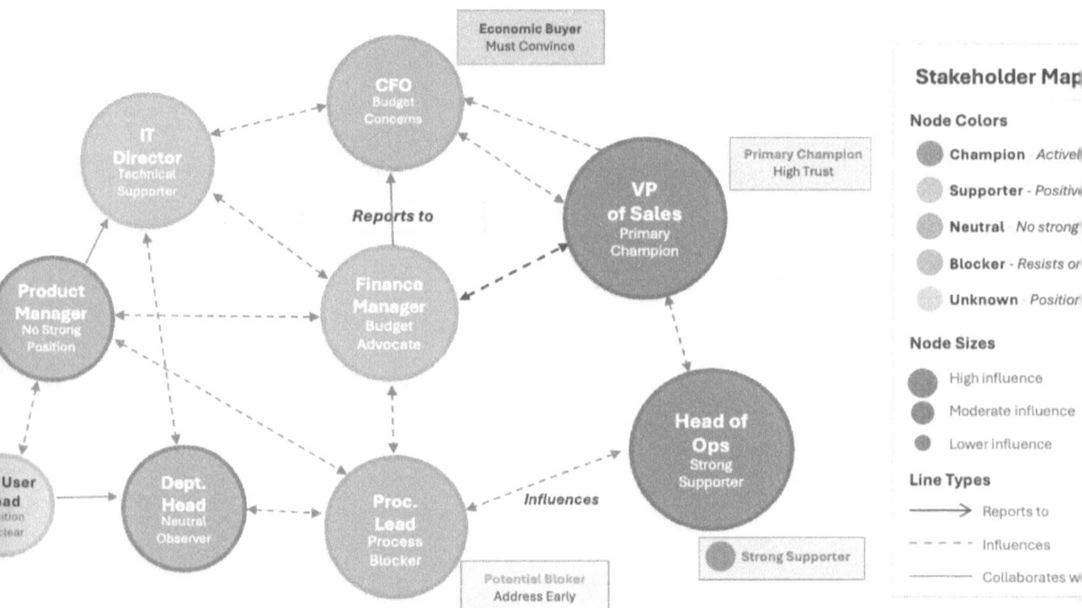

Figure 6.1: Stakeholder Influence Network Map for Complex B2B Sale

Identifying All Relevant Stakeholders

The first mistake most salespeople make is assuming they know who's involved based on initial conversations. The champion tells you, "I'm driving this decision" and you take that at face value, not realising there are five other people who can kill the deal.

Systematic stakeholder identification:

Economic buyer: Who controls the budget and has final spending authority? This isn't always the person who claims to have budget. It's whoever can approve expenditure without further escalation.

Technical buyer: Who validates that your solution will work in their environment? This is usually someone in IT, engineering, or operations who assesses feasibility, integration requirements, and technical risk.

User buyers: Who will use the solution day-to-day? Their adoption and satisfaction matter for long-term success, and they often have veto power if they resist strongly enough.

Coaches/Champions: Who actively advocates for your solution internally? These are individuals who've bought into your value proposition and will push for your selection even when you're not in the room.

Blockers: Who actively opposes your solution or has incentives to maintain status quo? This might be someone with a relationship with your competitor, someone who fears your solution threatens their role, or someone who simply has competing priorities.

Influencers: Who doesn't have formal decision authority, but whose opinion carries weight with decision-makers? This could be respected senior leaders, trusted advisors, or subject matter experts whose endorsement matters.

Validators: Who needs to sign off on specific aspects? Legal reviews contracts. Security assesses risk. Compliance ensures regulatory alignment. Procurement negotiates terms. Each can delay or block, even if they can't approve.

Most sales reps identify 3-4 of these stakeholders. Elite sellers identify 8-12. The difference isn't just thoroughness. It's understanding that in complex organisations, decision-making authority is distributed across multiple people, each with partial veto power.

Mapping Relationships and Influence

Once you've identified stakeholders, you need to understand how they relate to each other. Who influences whom? Who defers to whom? Who has political capital with whom?

Uncovering Hidden Relationships

Observe meeting dynamics: Who speaks first? Who speaks most? Who do others look to before answering questions? Who asks challenging questions versus supportive ones? These behaviours reveal actual influence regardless of titles.

Track communication patterns: Who gets copied on emails? Who forwards information to whom? Who requests input from whom? Communication flows reveal both formal and informal reporting structures.

Ask directly but carefully: "Who else typically weighs in on decisions like this?" "Whose opinion does [Executive Name] usually consult before finalising technology purchases?" "Are there other teams that need to sign off?"

Use your champion: Your internal coach can explain dynamics you can't see: "The CTO technically reports to the CEO, but on technology decisions he has full autonomy. However, he always runs major purchases by the CFO first. They have a long-standing relationship, and he values her financial perspective."

Use AI-powered relationship intelligence: Tools like LinkedIn Sales Navigator, Relationship Intelligence platforms, and some CRMs can map connections: who knows whom, who worked together previously, who shares board memberships or investors. These connections reveal influence pathways that aren't visible from org charts.

Assessing Individual Positions

For each stakeholder, determine:

Support level:

- Champion: Actively advocates for you
- Supporter: Positive but not actively pushing
- Neutral: No strong opinion yet
- Sceptic: Has concerns but not actively opposing
- Blocker: Actively resists your solution

Influence level:

- High: Their opinion heavily sways the decision
- Medium: They're consulted and their input matters
- Low: They're informed but don't significantly influence outcome

Engagement level:

- Engaged: Actively participating in evaluation
- Aware: Knows about it but not involved yet
- Unknown: Hasn't been brought into the conversation

Plot each stakeholder on these dimensions. This reveals critical gaps:

"We have a Champion with High Influence who is Engaged (good!), but the Economic Buyer is only Neutral with High Influence and Medium Engagement (problem!), and there's a Blocker with Medium Influence who is Highly Engaged (bigger problem!)."

This analysis tells you exactly where to invest effort: convert the Economic Buyer from Neutral to Supporter, neutralise the Blocker or reduce their influence, and maintain your Champion's enthusiasm while expanding engagement.

Multi-Threading: Building Relationships Across Stakeholders

biggest risk in complex sales is being single-threaded. Dependent on one relationship, one champion, one contact. When that person leaves, gets busy, loses influence, or simply can't drive the decision alone, your deal stalls or dies.

Multi-threading means developing relationships with multiple stakeholders so that your deal doesn't depend on any single person's success.

The Champion Paradox

Champions are essential. You need someone inside the organisation actively advocating for you. But champions are also dangerous if you're overly dependent on them.

Champion risks:

They leave the company: Happens more often than you'd think, especially in growth companies or during organisational change. When your champion leaves, you often leave with them.

They lose influence: Internal politics shift. Budgets get reallocated. Priorities change. Your champion's project gets deprioritised and they can no longer drive it.

They over-promise their influence: Many champions genuinely believe they have more authority than they do. They tell you "I'm driving this decision" when they're one voice among many.

They shield you from dissent: Champions want you to succeed, so they sometimes downplay internal opposition or hide concerns from other stakeholders, thinking they can handle it internally. Then you get blindsided.

They can't or won't introduce you to others: Sometimes champions want to maintain control of the relationship and resist connecting you with other stakeholders. This leaves you vulnerable.

The solution isn't to avoid champions. It's to build relationships beyond your champion so you're not solely dependent on them.

Strategic multi-threading

Effective multi-threading is systematic:

- **Start by understanding the stakeholder map:** You can't multi-thread until you know who all the relevant stakeholders are.
- **Identify the 3-5 most critical relationships:** You can't build deep relationships with 15 people. Identify the handful who are most critical. Usually the Economic Buyer, the Technical Buyer, a senior-level Champion, and any identified Blockers.

Develop a stakeholder engagement plan: For each critical stakeholder:

- Why do they matter to the decision?
- What do they care about personally and professionally?
- What concerns might they have?
- What value can you provide to them specifically?
- How can you get introduced or engage them?

Put to work your champion to facilitate introductions: The best way to multi-thread is through your champion: "It would be helpful to understand [CTO's] perspective on the technical architecture—could you introduce us?" Or: "I want to make sure [CFO] is comfortable with the financial model—should we schedule time to walk through it together?"

Most champions are happy to facilitate these connections because it helps them build internal buy-in.

Provide value in each interaction: When engaging new stakeholders, don't just pitch. Provide specific value relevant to them:

- For the CFO: Financial analysis, ROI models, risk mitigation
- For the CTO: Technical architecture, integration approach, security documentation
- For the COO: Operational impact, change management, implementation planning

Each stakeholder should feel you understand their specific concerns and are addressing them, not just giving the same pitch to everyone.

Maintain relationship equity: Multi-threading doesn't mean you stop relying on your champion. It means you're not solely dependent on them. Continue nurturing that relationship while building others.

Power Mapping: Understanding Who Really Decides

In complex organisations, formal authority (shown on org charts) rarely matches actual decision-making power. Understanding real power dynamics is essential for navigation.

Types of Power in Organisations

Formal authority power: This is title-based. The EVP has more authority than the Director. But formal authority alone doesn't guarantee influence.

Expert power: Someone with deep domain expertise whose opinion carries weight because they know more than others. The senior architect who's been with the company 15 years might have more influence on technology decisions than the new CTO.

Relationship power: Someone close to key executives. The Chief of Staff to the CEO often has more practical influence than many VPs because they have the CEO's ear daily.

Resource control power: Whoever controls budget, headcount, or other scarce resources has practical veto power even if they don't have decision authority. You might have the VP of Sales' enthusiastic support, but if the CFO controls spending and is sceptical, you're stuck.

Coalition power: Multiple stakeholders aligned create power that none have individually. If Sales, Marketing, and Customer Success all want your solution, their collective voice carries more weight than any single executive.

Political power: Some individuals have accumulated political capital through tenure, successful projects, or relationships. They can cash in that capital to drive decisions. Others have burned through their capital and struggle to get anything approved.

Mapping power means identifying who has which types of power and how they relate:

"The CTO has formal authority but is new to the company (low political capital). The VP of Engineering has been here 8 years and has the CEO's trust (high relationship power). On paper, the CTO decides. In practice, he'll defer to the VP of Engineering's recommendation."

Understanding this, you ensure the VP of Engineering is engaged and supportive, not just the CTO.

Identifying the True Economic Buyer

One of the most critical power questions is: who can approve spending?

This is often unclear because:

People claim authority they don't have: Directors say they control budget but need VP approval. VPs say they can approve but need CFO sign-off. Everyone wants to appear enabled.

Authority varies by deal size: Someone might have authority to approve $50K but need escalation for $100K. You don't know until you ask specifically: "What's your approval threshold? At what dollar amount does this need additional sign-off?"

Authority changes during the process: Budget might have been allocated to a director when the project was exploratory, but now that it's real, the CFO wants visibility. What started as a straightforward decision becomes multi-level approval.

Shared authority is common: In many organisations, no single person can approve alone. Sales and IT both need to agree. Finance and Operations both sign off. This diffused authority means you need coalition-building, not just one champion.

The best way to identify true economic buyer: ask directly and specifically:

"Help me understand the approval process. At $150K, does this require just your signature, or does it need additional approval? Who typically signs off on purchases at this level? Is there a formal approval hierarchy, or does it vary by project?"

Document the answers and validate them: "So if I understand correctly, you can approve up to $100K, but above that you need CFO sign-off, and all technology purchases over $50K also require CTO review for architecture approval. Is that accurate?"

Building Coalition: Orchestrating Multi-Stakeholder Buy-In

In complex sales, you rarely win by convincing one person. You win by orchestrating coalition. Getting multiple stakeholders aligned so that collective momentum overcomes individual hesitation and organisational inertia.

Coalition-Building Strategy

Start with your champion but expand systematically: Your champion is your foundation. They help you identify who else matters and facilitate introductions. But your goal is to build direct relationships and support beyond them.

Sequence engagement strategically: Don't approach all stakeholders simultaneously. Start with those most likely to support you, build their enthusiasm, then capitalise on their support when engaging sceptics.

For example:

- Solidify your champion (VP of Sales)
- Engage friendly stakeholders (Sales Operations Director)
- Bring in neutral but important stakeholders (VP of Marketing)
- Address technical stakeholders with supporting coalition (CTO, with Sales and Marketing support visible)
- Finally engage the Economic Buyer (CFO) with broad coalition support established

This sequence builds momentum. By the time you reach the CFO, you're not a vendor making a pitch. You're the solution that Sales, Marketing, and IT all support.

Customise value proposition for each stakeholder: Different stakeholders care about different things. Your coalition isn't built on one universal pitch, but on multiple tailored value propositions that align with each stakeholder's priorities.

Sales VP: "Your team will close deals 20% faster and forecast with 40% better accuracy."

Marketing VP: "You'll have visibility into how marketing campaigns translate to sales pipeline, proving marketing ROI."

CTO: "This integrates natively with your Salesforce instance. No custom development required, and we support your SSO architecture."

CFO: "3:1 ROI in first year, 180-day payback period, with pricing that scales with your revenue so there's no risk if growth slows."

Each stakeholder hears what matters to them, and collectively they all see value.

Create shared artifacts: Business cases, technical assessments, implementation plans, ROI models. These documents can be shared across stakeholders and become collective reference points that build shared understanding and alignment.

Instead of each stakeholder having their own private conversation with you, they all see the same materials and build shared context.

Facilitate multi-stakeholder conversations: Whenever possible, get multiple stakeholders together. Joint demos, group Q&A sessions, collaborative planning meetings. These create alignment because people hear each other's perspectives and can discuss trade-offs directly rather than sequentially through you.

"Let's get Sales, Marketing, and IT on a call together to discuss integration approach. That way we can address everyone's requirements at once and make sure we're aligned."

Make champions of the willing: Not everyone needs to be a strong advocate. Your goal is to convert fence-sitters into supporters, neutralise sceptics, and isolate blockers so they can't stop momentum.

You don't need unanimous enthusiasm. You need enough support that the decision can move forward despite some reservations.

Neutralising Blockers

In almost every complex deal, someone opposes your solution. Effective navigation means identifying blockers early and having strategies to address them.

Why People Block Deals

Understanding why someone blocks help you determine how to address it:

Competing priorities: They're not against you personally. They just have other initiatives they're prioritising and don't want resources diverted. Your solution competes with theirs for budget, attention, or implementation bandwidth.

Risk aversion: They've been burned before by vendors who over-promised and under-delivered. They're sceptical of anything new and prefer proven, low-risk approaches.

Relationship with competitor: They have an existing vendor relationship they're loyal to, either because it's working well or because they have personal connections.

Threat perception: Your solution threatens something they care about. Their budget, their authority, their team's role, or their established processes.

Legitimate concerns: Sometimes people block because they have real, valid concerns that haven't been adequately addressed. Security risks, integration complexity, adoption challenges.

Different strategy: They genuinely disagree about the right approach and believe an alternative solution or internal build is better.

Strategies for Neutralising Blockers

Address concerns directly: Don't avoid blockers or hope your champion handles them. Engage directly, understand their concerns, and address them specifically.

"I understand you have concerns about integration complexity. Can we walk through the technical architecture together so I can show you exactly how this works with your existing systems?"

Find common ground: Even blockers usually have goals you can align with. Find what they care about and show how your solution supports it.

"I know your priority is system stability and minimising technical risk. Our solution actually reduces risk by consolidating three-point solutions into one platform with better reliability."

Apply coalition: If you've built support from other stakeholders, use that momentum. The blocker might resist you individually but defer to broader organisational consensus.

"The Sales, Marketing, and Customer Success teams have all validated that this solves their core challenges. I want to make sure we're addressing your IT concerns as well so we can move forward with broad alignment."

Reduce their risk: Offer pilots, phased rollouts, or success guarantees that lower their personal risk if they support the decision.

"What if we started with a 30-day pilot with your Sales team? That way you can validate integration and performance in your environment before committing to enterprise deployment."

Bypass if necessary: If a blocker can't be won over, sometimes you need to go around them. This is politically delicate, but if they don't have actual veto power, you can succeed despite their opposition.

For your team, this requires understanding power mapping. If the blocker has limited actual influence, your coalition can override them. But if they have real power, you must address their concerns rather than bypass them.

When to Walk Away

Sometimes blockers can't be neutralised, and the organisational dynamics are too unfavourable to overcome. Recognising these early saves months of wasted effort.

Red flags that a deal is unwinnable:

- Multiple key stakeholders actively opposed with no path to convert them
- Political infighting so intense that no decisions get made
- Your champion has lost influence and can't drive consensus
- Economic buyer is sceptical and won't engage despite your best efforts
- Organisational chaos (merger, restructuring, leadership churn) freezes all decisions
- Incumbent vendor has relationships so strong that displacement is realistically impossible

Elite sellers know when to disqualify and move on. Not every deal is winnable, and spending six months fighting organisational dynamics you can't overcome is opportunity cost you'll never recover.

Managing the Decision Process

Beyond stakeholder relationships, complex deals require managing the actual decision-making process. The sequence of steps from interest to signature.

Understanding Formal and Informal Processes

Most organisations have formal purchasing processes: requirements definition, vendor evaluation, proposal submission, technical validation, legal review, contract negotiation, executive approval.

They also have informal processes: the conversations that happen in private meetings, the behind-the-scenes consensus-building, the political manoeuvring that determines which formal decisions get made.

Effective navigators manage both:

Formal process: Ensure you're meeting all official requirements, providing documentation on time, hitting milestones, addressing each stage's criteria. This is table stakes.

Informal process: Build relationships, create advocacy, address concerns before they become formal objections, orchestrate coalition before formal decision meetings. This is where deals are won or lost.

Critically, the mistake many sellers make: focusing exclusively on the formal process while ignoring the informal one.

They deliver great proposals, ace formal presentations, and provide all requested documentation. Then lose because they didn't build the behind-the-scenes support that determines what the formal decision will be.

Driving Momentum Through Decision Stages

Complex deals naturally lose momentum. Each stage transition—from discovery to demo to proposal to contract—creates opportunity for delays, new objections, or changed priorities.

Maintaining momentum requires:

Clear mutual commitment at each stage: Before ending any meeting, establish specific next steps with dates: "So we're aligned. You'll introduce me to the CTO by next Friday, we'll schedule a technical deep-dive the following week, and assuming that goes well, we'll aim for proposal submission by month-end. Does that timeline work?"

Making it easy to say yes: Remove friction at each stage. If they need technical documentation, provide it immediately. If they need references, have them ready. If they need budget justification, build the business case for them.

Creating urgency without pressure: Help them understand the cost of delay: "If we can get to contract by end of quarter, you'll go live before your busy season. Otherwise, we're looking at Q3 implementation during your peak, which is much more disruptive."

Anticipating and removing obstacles: Before each stage, ask: "What could slow this down? What concerns might come up? Who else needs to weigh in?" Then proactively address those issues before they become blockers.

Staying visible without being annoying: Regular touchpoints keep you top-of-mind: "Just checking in on the technical review. How's it going? Anything I can clarify?" You're helpful, not pushy.

AI-Powered Navigation Intelligence

Modern AI tools dramatically improve navigation capability by providing intelligence humans can't gather manually:

Relationship Mapping

LinkedIn Sales Navigator and specialized relationship intelligence platforms can map:

- Who knows whom (first, second, third-degree connections)
- Shared backgrounds (worked together previously, same university, etc.)
- Common connections (mutual contacts who can facilitate introductions)
- Job changes and career paths (understanding someone's trajectory predicts their influence)

This intelligence helps you: find warm introduction paths rather than cold outreach, understand which stakeholders already have relationships, identify who has history together (good or bad).

Organisational Intelligence

AI can track organisational changes that affect decision-making:

- Leadership changes (new executives often want to make their mark with new decisions)
- Reorganisations (shifting reporting structure changes power dynamics)
- Headcount changes (aggressive hiring signals growth priorities)
- Technology adoptions (what they're buying reveals strategy)

These signals tell you when to engage: "They just hired a new CRO who came from a company that used our platform. Perfect timing to reach out."

Deal Risk Prediction

Conversation intelligence platforms like Gong analyse patterns across thousands of deals to predict navigation risks:

"This deal has only one engaged stakeholder after 60 days. Historically, deals with single-threading at this stage have 14% close rate. Recommend expanding stakeholder engagement."

"The champion used the phrase 'I'll have to run this by leadership'. This language historically predicts 72% probability of stalled deal. Recommend directly engaging economic buyer."

"No mention of implementation timeline or budget approval process after three calls. Historically indicates lack of genuine urgency. Recommend explicit timeline and authority conversation."

Notice how thise alerts help you identify navigation gaps before they kill deals.

Competitive Intelligence

AI can monitor signals that competitors are engaged:

- Mentions of competitor names in calls (conversation intelligence catches this)
- Website visits to competitor sites (if you have web tracking)
- Intent data showing competitor research (from platforms like Bombora)
- Job postings or press releases mentioning technologies adjacent to competitors

Early awareness of competition lets you differentiate proactively rather than getting surprised late in the cycle.

Common Navigation Mistakes

Let's address the most common ways navigation goes wrong:

Mistake 1: Assuming the Org Chart Is Reality

Titles don't tell you who really influences decisions. The SVP might defer to the Director who has deep relationships. The person who looks junior might be a trusted advisor. Always validate actual power dynamics rather than assuming.

Mistake 2: Single-Threading

Relying on one champion is the biggest navigation risk. Always multi-thread, even when your champion is strong. People leave, lose influence, or simply can't drive the decision alone.

Mistake 3: Avoiding Difficult Conversations

When you sense problems—blocked access to stakeholders, concerns not being addressed, timeline slipping—address them directly rather than hoping they resolve themselves. They almost never do.

Mistake 4: Treating All Stakeholders the Same

Different people care about different things. Generic pitches don't build coalition. Tailor your approach to each stakeholder's priorities, concerns, and decision-making style.

Mistake 5: Neglecting the Informal Process

Winning on paper but losing in practice happens when you ace the formal evaluation but fail to build behind-the-scenes support. The decision is usually made informally before the formal approval process.

Putting Navigation into Practice

Let's bring this together with a practical example of how navigation works in a real deal:

Initial situation: You've been engaged by the VP of Sales at a 500-person SaaS company. She's enthusiastic about your sales automation platform. Discovery went great. The business case is solid.

Navigation assessment:

- Champion (VP of Sales): Strong support, high influence with CEO
- Economic buyer (CFO): Unknown position, high authority, not yet engaged
- Technical buyer (CTO): Sceptical of adding more tools, medium influence
- User buyers (Sales team): Mixed—some excited, some resistant to change
- Blocker (VP of Operations): Concerned about implementation disruption

Navigation strategy:

Week 1-2: Build coalition with friendly stakeholders

- Work with champion (VP Sales) to build detailed business case showing ROI
- Engage Sales Operations Director (reports to champion) to validate use cases
- Get early positive feedback from top-performing sales reps
- Create success metrics everyone can align on

Week 3-4: Address technical concerns

- Schedule technical deep dive with CTO and their engineering team
- Bring technical solutions engineer to address integration questions
- Provide detailed security documentation and architecture diagrams
- Get CTO to "neutral" or "supporter" position before engaging CFO

Week 5-6: Engage economic buyer with coalition support

- Champion schedules meeting with CFO, including VP Sales and CTO
- Present unified business case showing Sales, Operations, and IT alignment
- Emphasise financial model: 3:1 ROI, 180-day payback
- CFO gives conditional approval pending legal/procurement review

Week 7-8: Neutralise operations concerns

- Meet with VP of Operations to understand implementation concerns

- Propose phased rollout: pilot with 10 reps first, expand after validation
- Provide detailed change management plan and training approach
- Get VP Operations to "neutral" position. Won't advocate but won't block

Week 9-10: Guide legal and procurement

- Provide standard contract with company-favourable terms
- Address security questionnaire proactively with detailed responses
- Work through procurement's vendor management requirements
- Keep champion engaged so momentum doesn't stall in legal review

Week 11: Close with multi-stakeholder alignment

- Final approval meeting with CEO includes VP Sales, CFO, CTO
- CEO sees broad alignment and approves quickly
- Contracts signed, deal closed

Result: Deal that could have stalled or died (sceptical CTO, concerned Operations VP, unknown CFO) closed in 11 weeks because navigation was deliberate, systematic, and effective.

The Navigation Advantage

Here's what changes when you guide complex deals systematically:

Your win rates increase because you identify and address stakeholder concerns before they become deal-killers.

Your cycles shorten because you orchestrate the decision process actively rather than letting it drift.

Your deals stay alive because you're multi-threaded and not dependent on any single relationship.

Your forecasts improve because you understand actual decision dynamics, not just surface enthusiasm.

Your stress decreases because you're actively managing complexity rather than being surprised by it.

Most crucially, you move from hoping deals close to making them close through deliberate orchestration of organisational dynamics.

In the next chapter, we'll explore Tailor—how to personalise your approach at scale across all these stakeholders without sacrificing quality or burning yourself out. Because navigation tells you who to engage and how to orchestrate them, but tailoring determines what you say to each person to land and drive action. In the next chapter, we'll explore what you actually say to each person. How to tailor your approach at scale without burning yourself out.

CHAPTER OUTCOME

- ☐ A stakeholder influence map for a current opportunity (not just names. Power + influence)
- ☐ A champion plan (who, why they care, what they need to win internally)
- ☐ A multi-threading plan (who you will engage, in what order, with what message)
- ☐ A blocker plan (likely blockers + how you'll neutralise them with facts and alignment)
- ☐ A "deal orchestration" timeline: milestones, decision gates, and internal buying steps
- ☐ A process for managing informal decision dynamics (not just the formal process)
- ☐ A set of signals you'll monitor for deal risk and competitive moves

CHAPTER 7

Tailor – Personalisation at Scale

By the end of this chapter, you will be able to:

- Choose the right level of personalisation (from segment-based to stakeholder-specific) without burning out.
- Use AI for content generation while maintaining strategic oversight and authenticity.
- Personalise across multiple stakeholders with consistent messaging and role-relevant value framing.
- Build dynamic proposal generation and apply advanced personalisation where it counts.
- Measure effectiveness and avoid common traps (personalisation theatre, inaccuracies, inconsistency, forgetting the human).

Delivering customised experiences without burning out

A few years ago, I was coaching a sales rep who was struggling with a common problem: she knew personalisation mattered, but she didn't have time to do it properly. She was managing 40 active opportunities, each with multiple stakeholders, and every piece of advice she'd ever received said "personalise everything."

So, she tried. She spent hours researching each prospect's LinkedIn profile, reading their blog posts, crafting custom emails referencing their interests. She built unique presentations for every demo. She wrote proposals from scratch for each opportunity, carefully tailoring the language to what she'd learned about their business.

The result? She was working 70-hour weeks and still missing quota. The personalisation was good—prospects appreciated the effort—but she couldn't scale it. She spent so much time customising that she didn't have enough time selling. And ironically, the proposals she spent 6 hours perfecting didn't close at meaningfully higher rates than colleagues who used templates and spent 30 minutes customising them.

She'd fallen into what I call the "personalisation paradox": everyone wants personalised experiences, but true personalisation doesn't scale without tap into.

This is what the Tailor stage of the QUANTUM Framework solves. You've qualified the right prospects, understood them deeply, anticipated their concerns, and navigated their organisational complexity. Now you need to communicate with them in ways that feel genuinely personalised. But you need to do it across dozens of prospects, each with multiple stakeholders, without spending your entire life writing custom content.

The answer isn't choosing between personalisation and scale. It's using AI to deliver authentic personalisation at scale by automating the automatable while preserving human judgment on what matters.

The Personalisation Spectrum

Before we dive into how to personalise at scale, let's establish that personalisation isn't binary. It exists on a spectrum from completely generic to completely customised. Understanding this spectrum helps you make intelligent choices about where to invest effort.

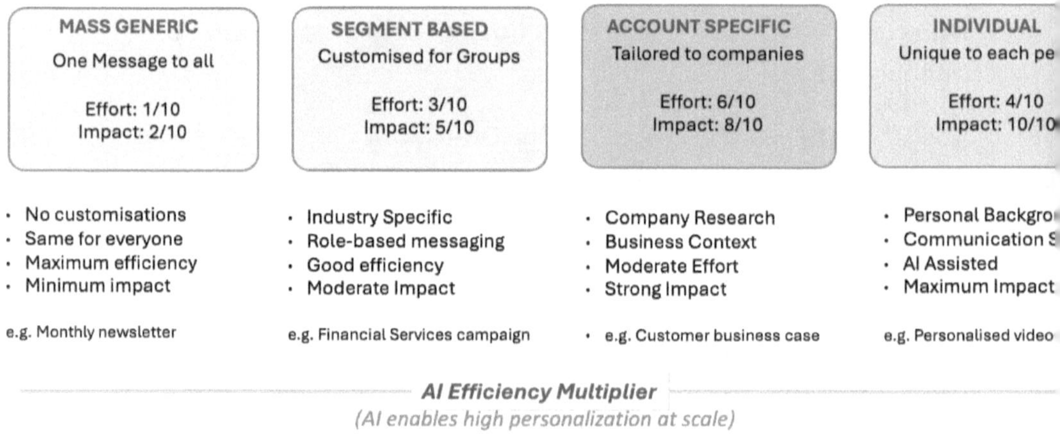

Figure 7.1: The Personalisation Spectrum—Balancing Effort, Impact, and Scale. From Generic Broadcasts to Individual Customisation

Level 1: Mass Generic

This is one-size-fits-all communication. The same message goes to everyone regardless of who they are, what industry they're in, or what problems they face.

Examples:

- Newsletter blasts: "Check out our new feature release!"
- Generic cold emails: "We help companies improve sales productivity."
- Standard demo: Same presentation for every prospect
- Templated proposals: Replace company name, everything else identical

When this works: Never, really. Even early-stage awareness campaigns benefit from segmentation. Mass generic is what lazy marketers do, not what effective sellers do.

Effort required: Minimal—write once, send to everyone.

Impact: Very low. Recipients immediately recognise its generic and mostly ignore it.

Level 2: Segment-Based

As a result, this is personalisation by category. You customise for groups with shared characteristics. Industry, company size, role, use case.

Examples:

- Industry-specific emails: "Here's how financial services companies use our platform…"
- Role-based content: Different messaging for CFOs versus VPs of Sales
- Use case presentations: Demo focused on their specific workflow

- Vertical proposals: Healthcare proposal template, Tech company proposal template

When this works: This is the minimum acceptable personalisation for B2B sales. Prospects expect you to understand their industry and role at least at a categorical level.

Effort required: Moderate—create templates for major segments, reuse across prospects in those segments.

Impact: Moderate. Better than generic but still feels somewhat impersonal since prospects know you're using this same approach for others like them.

Level 3: Account-Specific

This is customisation at the company level. You tailor your approach to this specific organisation's situation, challenges, and context.

Examples:

- Company research in outreach: "I noticed you just opened offices in Europe..."
- Business-specific demos: Using their workflow, their terminology, their use cases
- Customised business cases: ROI model based on their specific metrics and situation
- Company-tailored proposals: References to their strategy, competitors, challenges

When this works: This is what separates good sellers from average ones. Prospects feel understood when you clearly know their specific business context, not just their industry generally.

Effort required: Significant—requires research, customisation, thoughtful application of insights.

Impact: High. Prospects engage more deeply when they feel you understand their specific situation.

Level 4: Individual-Personalised

This is customisation at the stakeholder level. You tailor not just to the company, but to each individual person based on their role, priorities, communication style, and personal motivations.

Examples:

- Personalised outreach: References to their specific background, interests, recent posts
- Stakeholder-specific value propositions: Different messaging for CFO versus CTO versus VP Sales
- Communication style matching: Analytical detail for some, high-level vision for others
- Personal connection points: Shared interests, mutual connections, common experiences

When this works: This is what elite sellers do for key stakeholders in important deals. You can't do this for everyone, but for the people who most influence major decisions, individual personalisation creates disproportionate impact.

Effort required: Very significant—deep research, careful customisation, thoughtful relationship building.

Impact: Very high. When someone feels you've taken time to understand them as an individual, not just a title, trust and engagement accelerate dramatically.

The Personalisation Paradox

Here's the challenge: the most effective personalisation (individual-level) is also the most time-intensive. If you tried to deliver Level 4 personalisation to every stakeholder in every deal, you'd have time for maybe 3-5 opportunities total.

But if you only do Level 2 (segment-based), you'll lose to competitors who are doing Level 3 and 4 for key accounts and stakeholders.

Consider how the solution is strategic personalisation. Delivering the right level of personalisation to the right people at the right time, using AI to scale where possible while investing human effort where it matters most.

The Personalisation Decision Matrix

Not every interaction deserves the same level of personalisation. Here's how to decide:

Use Level 4 (Individual-Personalised) for:

- Economic buyers in high-value deals
- Key champions who drive deals forward
- Known blockers you need to convert
- First outreach to strategic target accounts
- Final presentation/negotiation with C-level executives

Use Level 3 (Account-Specific) for:

- All stakeholders in qualified opportunities
- Proposals and business cases
- Technical deep-dives and demos
- Important follow-up communications

Use Level 2 (Segment-Based) for:

- Initial outreach to prospects not yet qualified
- Nurture campaigns for pipeline development
- Follow-up to less critical stakeholders
- Educational content distribution

Never use Level 1 (Mass Generic):

- Even the least important communication benefits from basic segmentation

For your team, this matrix means you're delivering high-touch personalisation where it matters (major deals, key stakeholders) while efficiently handling everything else with segment-based approaches.

AI-Powered Content Generation

The gamechanger in personalisation at scale is AI's ability to generate customised content that would take humans hours to write, in seconds.

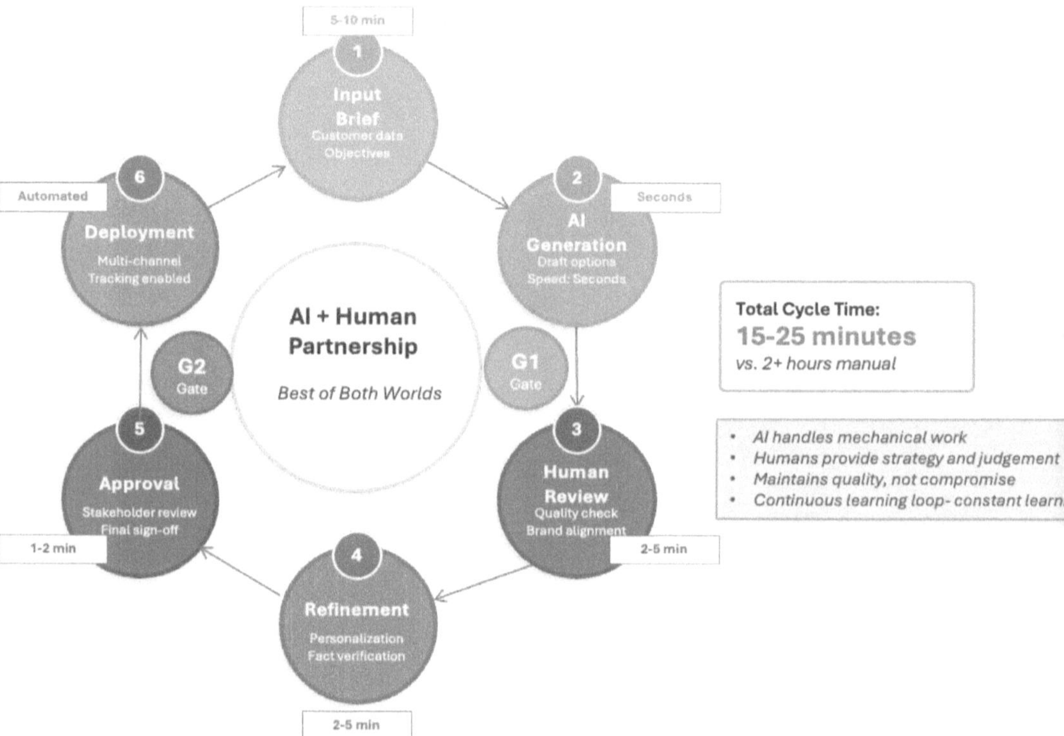

Figure 7.2: AI-Assisted Content Generation with Human-in-the-Loop Quality Control

Modern large language models (like ChatGPT, Claude, or specialized sales AI tools) can:

- Write emails tailored to specific prospects based on research about their company
- Generate proposals incorporating specific business context and requirements
- Create presentations adapted to audience, industry, and use case
- Customise messaging to match communication styles and preferences
- Draft follow-up content referencing previous conversations

The key is understanding how to use AI effectively. Not as a replacement for human judgment, but as a multiplier of human capability.

The AI Content Generation Workflow

Here's the systematic process for AI-assisted personalisation:

Step 1: Gather Context

Before AI can generate relevant content, it needs context:

- Who is this for? (Company, role, individual)
- What do we know about them? (Research, previous interactions, pain points)
- What's the objective? (First outreach, follow-up, proposal, demo invitation)
- What tone/style? (Formal, conversational, technical, high-level)
- What constraints? (Length, format, required elements)

Modern CRMs and sales engagement platforms can aggregate this context automatically, pulling from:

- Account records (firmographic data, opportunity details, notes)

- Conversation intelligence (call transcripts, meeting summaries)
- Email history (previous exchanges, engagement data)
- Research tools (company information, stakeholder profiles)

Step 2: Generate Initial Draft with AI

Provide the AI with rich context and clear instructions:

"Write a personalised email to Sarah Chen, CFO at TechStars Inc (250-person B2B SaaS company, Series B funded, growing 40% YoY). She's expressed concern about sales forecast accuracy in our last call. Reference her background in investment banking and her focus on operational efficiency. Propose a 30-minute follow-up to walk through our ROI model. Tone should be professional but personable, 150-200 words."

Looking closer, the AI generates a draft that incorporates this context, typically better than a generic template but not yet perfect.

Step 3: Human Review and Refinement

This is the critical step many people skip. And why AI-generated content sometimes feels robotic or off-target.

A human reviews the draft and refines:

- **Factual accuracy:** Does everything stated align with what we know?
- **Tone calibration:** Does this sound like how I communicate?
- **Strategic emphasis:** Are we highlighting the right points for this stakeholder?
- **Personalisation authenticity:** Do the personal touches feel genuine or forced?
- **Call to action clarity:** Is it clear what we want them to do next?

Good refinement takes 2-5 minutes versus 20-30 minutes to write from scratch. You're editing, not creating. A massive efficiency gain.

Step 4: Approval and Deployment

For important communications (proposals, executive outreach, key stakeholder engagement), a senior rep or manager reviews before sending. For routine communications (follow-ups, nurture emails, scheduling), reps can approve and send directly.

Step 5: Learning and Optimisation

Track which AI-generated content performs well (opens, responses, conversions) and which doesn't. Feed this data back to improve prompts and templates.

Over time, the AI learns what works for your specific audience, improving quality automatically.

What AI Does Well in Personalisation

Incorporating factual context: AI excels at taking research about a company—their industry, size, recent news, tech stack—and weaving it naturally into communications. "I noticed TechStars recently expanded into European markets and is scaling your sales team quickly. Most companies at your stage struggle with..."

Adapting tone and style: Give AI examples of your writing style and it can match it remarkably well, making generated content sound like you rather than generic corporate-speak.

Scaling repetitive customisation: If you're sending similar messages to 20 prospects in financial services, AI can customise each for the specific company while maintaining consistent messaging about your value for financial services companies.

Generating variations: Need three different subject lines to A/B test? Five different opening paragraphs? AI generates variations instantly, letting you test what works without spending hours writing alternatives.

Maintaining consistency: When personalising for multiple stakeholders, AI helps ensure your messaging is consistent. You're not accidentally contradicting yourself in the email to the CFO versus the CTO.

What AI Doesn't Do Well (Yet)

Deep strategic thinking: AI can't determine *whether* you should send an email, *who* you should focus on, or *what* your core strategy should be. Those remain human judgment calls.

Genuine relationship building: AI can draft personalised messages, but it can't replace the human connection built through authentic conversation, empathy, and shared experience.

Reading subtle context: AI might miss details from a conversation. That the prospect seemed hesitant about something, that a particular topic is politically sensitive, that a certain approach would backfire given interpersonal dynamics.

Creativity and insight: AI generates plausible, professional content. It rarely generates truly creative approaches or breakthrough insights that surprise and delight prospects.

Ethical judgment: AI doesn't know when personalisation crosses into manipulation or when a prospect's privacy should be respected versus applied.

The sweet spot is AI handling mechanical personalisation (company research, segment customisation, tone matching) while humans provide strategic direction, relationship judgment, and final polish.

Personalising Across Multiple Stakeholders

In complex B2B sales, you're not personalising for one person. You're customising your approach for 5-10 stakeholders, each with different priorities, communication styles, and concerns.

This is where AI-powered personalisation becomes essential. Manually creating customised content for each stakeholder in each deal is impossibly time-consuming. AI makes it practical.

The Multi-Stakeholder Personalisation Strategy

Step 1: Map stakeholder profiles

For each key stakeholder, document:

- Role and responsibilities
- Personal background and expertise
- Communication style (data-driven vs. visionary, detail vs. high-level)
- Primary concerns and priorities
- Success metrics they're evaluated on
- Relationship to other stakeholders

Modern CRMs can store this as structured data, making it available to AI for content generation.

Step 2: Create stakeholder-specific value propositions

What does your solution mean to each person?

CFO: "3:1 ROI in first year, 180-day payback, pricing scales with revenue so investment risk is minimised"

VP of Sales: "Reps close deals 20% faster, forecast accuracy improves 40%, top performers' best practices scale across team"

CTO: "Native Salesforce integration, enterprise-grade security, SSO support, proven at 500+ companies with 99.9% uptime"

VP of Operations: "Phased rollout minimizes disruption, complete training included, dedicated customer success manager"

AI can incorporate these value propositions into customised communications for each stakeholder automatically.

Step 3: Generate stakeholder-specific content

Instead of one generic proposal, create customised sections for each key stakeholder:

- **Executive summary for CEO:** High-level strategic impact
- **Financial analysis for CFO:** Detailed ROI model with sensitivity analysis
- **Technical architecture for CTO:** Integration approach, security documentation, scalability
- **Implementation plan for Operations:** Timeline, resources, change management
- **User adoption plan for Sales VP:** Training, rollout strategy, success metrics

AI can generate these sections based on templates and stakeholder profiles, then humans review and refine for accuracy and emphasis.

Step 4: Coordinate messaging

Ensure your personalised messages to different stakeholders are consistent. You're not promising different things to different people. AI helps by maintaining a central source of truth and flagging potential contradictions.

Step 5: Track engagement by stakeholder

Monitor which stakeholders are engaging with your content, what they're focusing on, and where concerns might be emerging. This intelligence informs your follow-up personalisation.

Practical Example: Multi-Stakeholder Email Sequence

You're following up after an initial demo with a 300-person SaaS company. You need to engage four stakeholders. Here's how AI-powered personalisation works:

Email to VP of Sales (Champion): AI-generated draft incorporates:

- Reference to their specific pain point mentioned in demo (forecast accuracy)
- Relevant case study from similar-sized SaaS company
- Concrete metrics (20% faster sales cycles, 40% forecast improvement)
- Call to action: next steps to build internal business case

Human refinement:

- Adds personal note about their recent LinkedIn post
- Adjusts tone to match their communication style from previous exchanges
- Sharpens the call to action based on where this deal really is

Email to CFO (Economic Buyer): AI-generated draft incorporates:

- Financial focus appropriate for CFO audience
- ROI calculation specific to their company size and sales team
- Risk mitigation points (scalable pricing, no large upfront commitment)
- Call to action: 15-minute financial review meeting

Human refinement:

- Emphasises points that resonated in demo (payback period)
- Addresses specific concern CFO raised about implementation cost
- Adjusts number formatting to match their style (noticed CFO prefers tables to prose)

Email to CTO (Technical Stakeholder): AI-generated draft incorporates:

- Technical language appropriate for engineering audience
- Specific integration points with their known tech stack (Salesforce, Slack)
- Security and compliance documentation
- Call to action: technical deep-dive session

Human refinement:

- References technical question CTO asked in demo
- Adds link to architecture documentation CTO specifically requested
- Adjusts to address CTO's preference for detailed written specs before meetings

Email to VP of Operations (Concerned about Implementation): AI-generated draft incorporates:

- Implementation focus addressing disruption concerns
- Case study from company with similar operational complexity
- Phased rollout approach
- Call to action: implementation planning conversation

Human refinement:

- Emphasises change management support (picked up on VP's concern about team adoption)
- Adds specific timeline reflecting their business constraints (busy season coming)
- Adjusts tone to be more reassuring (sensed this person is risk-averse)

Result: Four personalised emails that address each stakeholder's specific priorities, generated in 15 minutes total instead of 2+ hours writing from scratch. Each feel personally crafted because the human refinement adds authentic touches AI couldn't generate.

Dynamic Proposal Generation

Proposals are where personalisation traditionally takes the most time. And where AI can create the most apply.

Traditional approach: Start with last quarter's proposal, manually find-and-replace company names, adjust some business context, update pricing, hope you didn't miss anything. Takes 4-6 hours. Still feels somewhat generic.

AI-powered approach: Generate proposal dynamically from structured data, incorporating specific context automatically. Human reviews and refines specific sections. Takes 30-60 minutes. Feels customised because it is.

The Dynamic Proposal System

Component 1: Structured Content Library

Instead of monolithic proposal documents, break content into modular components:

- Executive summary templates (by industry, company size, use case)
- Problem statements (by vertical, situation, pain point)
- Solution descriptions (by product, feature set, configuration)
- Case studies (by industry, company size, use case, outcome)
- Technical architecture (by integration scenario, platform, requirements)
- Implementation plans (by timeline, complexity, resource needs)
- Pricing and packaging (by deal size, contract terms, add-ons)
- Terms and conditions (by deal type, geography, special requirements)

Each component exists in multiple variations optimised for different contexts.

Component 2: Intelligent Assembly Rules

AI selects which components to include based on:

- Prospect ICP characteristics (industry, size, stage)
- Opportunity details (deal size, timeline, stakeholders)
- Pain points and requirements identified in discovery
- Competitive context (what alternatives are they evaluating)
- Previous successful proposals for similar companies

"For a 200-person financial services company evaluating both us and Competitor X, with primary focus on forecast accuracy and integration with Salesforce, include: Executive summary template 3, Problem statement FS-2, Solution description focusing on predictive analytics, Case study from Regional Bank success, Technical architecture for Salesforce native integration, Implementation plan for phased rollout, Pricing tier appropriate for 200-person company, Terms including security addendum required for financial services."

Component 3: Dynamic Content Customisation

Within each selected component, AI customises:

- Company name and specifics throughout
- Industry-specific language and terminology
- Reference to specific pain points from conversations
- Metrics and outcomes relevant to their situation
- Stakeholder names and roles
- Timeline reflecting their constraints

Put simply, this isn't just find-and-replace. It's contextual adaptation. The problem statement discusses their specific challenges using language from your conversations. The case study emphasises outcomes most relevant to their priorities. The implementation timeline accounts for their busy season and resource constraints.

Component 4: Human Review and Refinement

AI generates a 90% complete proposal in 2 minutes. Human reviews and refines:

- Validates factual accuracy (did we capture their situation correctly?)
- Adjusts emphasis (are we highlighting what matters most to this buyer?)
- Adds personal touches (specific references from conversations)
- Ensures quality and polish (does this reflect our brand well?)
- Confirms strategic alignment (does this position us optimally vs. competition?)

This review takes 30-60 minutes versus 4-6 hours to create from scratch.

Component 5: Version Control and Collaboration

Multiple stakeholders might need to review. Sales, sales engineering, legal, executive sponsor. AI-powered proposal systems enable:

- Tracked changes so everyone sees what's being modified
- Comments and feedback without breaking the document
- Approval workflows to ensure sign-off before sending
- Version history in case you need to revert changes

Advanced Proposal Personalisation

Beyond basic customisation, AI enables sophisticated proposal techniques:

Stakeholder-specific sections: Generate different sections for different readers. The CFO gets detailed financial analysis. The CTO gets technical architecture deep-dive. The CEO gets strategic summary. All in one document, with personalised cover letters for each stakeholder highlighting what's most relevant to them.

Interactive proposals: Some platforms generate web-based proposals where content dynamically adapts based on who's viewing and what they click on. CFO clicks on ROI—it expands with full financial model. CTO clicks on integration—technical documentation appears. Engagement tracked to see what each stakeholder focused on.

Competitive positioning: If you know they're evaluating Competitor X, AI can incorporate comparison content highlighting your differentiation. If they're considering internal build, content addresses that alternative specifically.

Pricing optimisation: AI can suggest pricing based on similar deals, helping you optimise between competitive position and margin. "Companies this size in financial services typically sign at $150-180K. Recommend starting at $175K with flexibility to $165K if needed for competitive reasons."

Maintaining Authenticity at Scale

The biggest concern about AI-powered personalisation is whether it feels authentic or manipulative. Here's the key insight: personalisation is only valuable if it's genuine.

Authentic personalisation means:

You're tailoring to real information, not pretending to know things you don't: If you reference someone's background or interests, it's because you genuinely researched it and it's relevant, not because you're name-dropping to fake connection.

You're providing value, not just appearing to care: Personalised content should help the prospect. Giving them relevant insights, addressing their real concerns, saving them time—not just be flattery designed to manipulate.

You're consistent across touchpoints: If you tell the CFO one thing and the CTO something different, that's not personalisation, it's deception.

Your messaging should be adapted for audience but fundamentally consistent in facts and positioning.

You're transparent about using AI when appropriate: Most prospects don't care whether content was drafted by AI or human, as long as it's relevant and accurate. You don't need to disclose every use of AI, but you shouldn't pretend something is hand-crafted when it's templated.

You deliver on what you promise: If your personalised outreach promises specific value or insights, you need to deliver that in follow-up. Personalisation that's just bait-and-switch creates distrust.

The Authenticity Checklist

Before sending personalised content, ask:

- **Is this information accurate?** Have we correctly understood their situation?
- **Is this relevant?** Does this personalisation add value, or is it just showing off research?
- **Would this person feel respected if they knew how we created it?** Or would they feel manipulated?
- **Are we consistent?** Have we said the same thing to other stakeholders, adapted appropriately?
- **Can we back up what we're claiming?** If we say we understand their challenges, can we demonstrate that in conversation?

If you can answer yes to all of these, your personalisation is authentic, regardless of whether AI helped generate it.

Personalisation Across Communication Channels

Personalisation isn't just about email and proposals. It extends across every touchpoint:

Sales Calls and Meetings

Before meeting:

- AI generates briefing document with relevant context about attendees, their company, recent developments
- Suggests talking points based on their known priorities
- Flags potential concerns based on similar conversations

During meeting:

- Conversation intelligence captures what's discussed in real-time
- AI can surface relevant content (case studies, technical docs) mid-meeting to share
- Suggests follow-up questions based on patterns in successful calls

After meeting:

- AI generates meeting summary and action items
- Drafts personalised follow-up email referencing specific discussion points
- Updates CRM with relevant insights captured

Presentations and Demos

Dynamic demos: Instead of fixed presentation deck, modular content that adapts based on:

- Who's in the room (executive vs. technical audience)
- What you learned in discovery
- Which features map to their requirements
- Time available (30-min executive briefing vs. 2-hour technical deep-dive)

Personalised demo environments: For technical demonstrations, pre-configure demo with:

- Their company branding

- Sample data reflecting their use case
- Workflows matching their processes
- Integrations with their actual tech stack (in demo environment)

This makes the solution feel real and applicable, not abstract and generic.

Content Sharing and Follow-Up

Intelligent content recommendations: Based on prospect's profile, stage, and concerns, AI suggests which case studies, whitepapers, videos, or blog posts to share. Not random content. Specifically relevant based on similar buyers' engagement patterns.

Personalised content delivery: When sharing content, add context: "Given your focus on implementation timeline, this case study from a company your size shows how the phased rollout worked in practice. The key insight starts on page 4."

Consider that this is more valuable than just sending a link.

Contract Negotiation

Personalised terms: Based on deal size, buyer priorities, competitive context, and standard terms, AI suggests contract structure:

- Payment terms aligned with their budget cycle
- SLAs matching their risk tolerance
- Pricing structure optimised for their use case (per-user vs. flat fee vs. usage-based)

Redline analysis: When prospects request contract changes, AI flags:

- Which changes are standard/acceptable vs. problematic
- What similar companies accepted in previous deals
- Risk implications of specific terms

This speeds negotiation by avoiding unnecessary back-and-forth on standard terms while focusing discussion on genuinely material items.

Measuring Personalisation Effectiveness

How do you know if personalisation is working? Track these metrics:

Engagement metrics:

- Email open rates (personalised vs. generic)
- Response rates
- Meeting acceptance rates
- Content engagement (time spent, pages viewed)
- Proposal views and forwards

Conversion metrics:

- Qualification rate (how many initial contacts become qualified opportunities)
- Progression rate (how quickly do opportunities advance through stages)
- Win rate (does personalisation correlate with closed deals)
- Deal size (do personalised approaches command better pricing)

Efficiency metrics:

- Time spent per communication (AI-assisted vs. fully manual)
- Communications sent per rep per day
- Number of stakeholders effectively engaged per deal

Quality metrics:

- Prospect feedback (do they comment on relevance and value)
- Internal team assessment (does content meet quality standards)
- Sales leadership review (are reps personalising appropriately)

Pay attention to the goal isn't perfect personalisation. Its optimal personalisation given time constraints. You're looking for the sweet spot where additional personalisation effort produces diminishing returns.

Common Personalisation Mistakes

Let's address how personalisation goes wrong:

Mistake 1: Personalisation Theatre

Mentioning someone's college or shared hometown just to show you looked at their profile isn't personalisation. It's performance. Unless there's genuine relevance (you're both alumni and there's a real connection point), it feels forced and manipulative.

Better: Only reference personal information when it's relevant to your conversation or creates genuine connection.

Mistake 2: Over-Personalisation

Making every single communication completely unique is unsustainable and unnecessary. Save deep personalisation for high-impact moments (first outreach, key stakeholder engagement, final presentations).

Better: Use segment-based personalisation for routine touchpoints and invest in individual personalisation strategically.

Mistake 3: Inaccurate Personalisation

Getting details wrong is worse than being generic. If you reference someone's background or company situation incorrectly, you've signalled you don't understand them. Opposite of your goal.

Better: Validate facts before incorporating them. When uncertain, stay more general rather than risk inaccuracy.

Mistake 4: Inconsistent Personalisation

Sending a highly personalised initial email then following up with generic templates is jarring. Set expectations you can sustain.

Better: Maintain consistent level of personalisation appropriate for the relationship stage and deal importance.

Mistake 5: Forgetting the Human

Relying entirely on AI-generated content without human review leads to robotic, occasionally inaccurate, sometimes tone-deaf communications.

Better: Always have humans review AI-generated content for important communications. AI drafts, humans refine.

The Personalisation Capability Stack

To execute personalisation at scale effectively, you need the right tools:

Research and intelligence tools:
- LinkedIn Sales Navigator (professional background, connections)
- ZoomInfo, Apollo (firmographic and contact data)
- Built With, Datanyze (technology stack intelligence)
- News aggregators and Google Alerts (company developments)

Content generation tools:
- ChatGPT, Claude (general-purpose content generation)
- Jasper, Copy.ai (marketing-focused content generation)
- Native CRM AI (Salesforce Einstein, HubSpot AI)
- Specialized sales AI (Lavender, Regie.ai)

Proposal and presentation tools:
- Panadol, Proposify (dynamic proposal generation)
- High spot, Seismic (sales content management with personalisation)
- Beautiful.ai, Pitch (AI-assisted presentation creation)

Email personalisation tools:

- Outreach, SalesLoft (sales engagement with AI assist)
- Lavender (email coaching and personalisation)
- Crystal (personality insights for communication style matching)

Conversation and meeting tools:

- Gong, Chorus.ai (conversation intelligence)
- Zoom, Microsoft Teams (meeting platforms with AI transcription)
- Otter.ai (meeting notes and summaries)

The most effective approach: integrated stack where tools share data so context flows naturally, and AI can personalise across all touchpoints, not just individual tools.

Building Your Personalisation Workflow

Here's a practical daily workflow for personalisation at scale:

Morning (30 minutes):

- Review AI-generated prioritisation of accounts/opportunities
- Check personalised recommendations for today's outreach
- Review and approve AI-drafted follow-ups from yesterday's calls
- Customise 2-3 high-priority communications for key stakeholders

Throughout day:

- Use AI-assisted talking points for calls and meetings
- Let conversation intelligence capture meeting details
- Share AI-recommended content with prospects based on conversations
- Quick review/approve routine AI-generated emails (2-3 minutes each)

End of day (20 minutes):

- Review AI-generated meeting summaries and action items

- Approve tomorrow's outreach queue (AI has drafted, you review)
- Flag any needed deep personalisation for tomorrow
- Check engagement metrics on today's personalised communications

Weekly (1 hour):
- Review personalisation effectiveness metrics
- Refine AI prompts and templates based on what's working
- Deep-customise key communications for next week's strategic opportunities
- Update stakeholder profiles in CRM to improve AI context

Result: Delivering account-specific and individual-personalised content at scale without burning out.

The Personalisation Advantage

Here's what changes when you personalise systematically with AI:

Your response rates increase because prospects engage with relevant, tailored content rather than generic pitches.

Your relationships deepen because stakeholders feel understood as individuals, not just titles.

Your win rates improve because personalised value propositions hit home more strongly than generic ones.

Your efficiency increases because AI handles mechanical personalisation while you focus on strategy and relationship building.

Your differentiation strengthens because most competitors still send generic content. Your personalisation stands out.

Most crucially, you escape the personalisation paradox. You deliver genuine customisation at scale without sacrificing your evenings and weekends.

In the next chapter, we'll explore Unify. How to integrate all these personalisation capabilities, along with the rest of your AI-powered sales stack, into a smooth ecosystem where intelligence flows and compounds across every touchpoint.

Because personalisation is only as good as the intelligence behind it, and that intelligence needs to be unified across your entire operation.

CHAPTER OUTCOME

☐ Your chosen personalisation levels (segment / account / stakeholder) and when to use each

☐ A messaging framework that stays consistent across stakeholders (core narrative + role-based angles)

☐ A content workflow (AI draft → human review → accuracy check → tone check → deliver)

☐ A "personalisation accuracy" rule-set (what must be verified before sending)

☐ 5–10 reusable prompts for emails, call prep, follow-ups, proposals, and exec summaries

☐ A proposal structure you can assemble quickly (sections, proof points, case examples)

☐ A way to measure if personalisation is working (reply rate, meeting rate, velocity, quality of engagement)

CHAPTER 8

Unify – Integration and Orchestration

By the end of this chapter, you will be able to:

- Diagnose the integration crisis (tool sprawl, data silos, duplicated work, missed signals).
- Understand the four-layer sales technology model (data foundation → AI layer → applications → UI).
- Choose an integration architecture (CRM hub-and-spoke, iPaaS, warehouse/reverse ETL) based on your reality.
- Build a unified sales stack using a practical 5-step build sequence (audit → priorities → approach → governance → intelligence).
- Avoid common pitfalls (over-engineering, integrating bad data, UX neglect, fragile dependencies, "set and forget").

Building a sales ecosystem where intelligence flows cleanly

Three years ago, I watched a sales team at a fast-growing SaaS company slowly drown in their own success. They'd invested heavily in best-of-breed tools: Salesforce for CRM, Outreach for sales engagement, Gong for conversation intelligence, ZoomInfo for prospecting data, Tableau for analytics, and half a dozen other specialized solutions. Each tool was excellent at its specific function. The problem wasn't the tools. It was that they didn't talk to each other.

The result was predictable chaos. A prospect would fill out a web form, triggering a lead in Salesforce. But that lead data didn't automatically flow to Outreach, so someone had to manually add it to their cadence. When a sales rep had a great discovery call captured in Gong, the insights didn't update the opportunity in Salesforce. The rep had to remember to add notes themselves. When deals closed, the handoff to customer success involved copying information between three different systems.

In real terms, the team was spending 30% of their time on administrative work. Updating records, copying data between systems, reconciling information that should have been automatically synchronised. And despite all this effort, their pipeline data was still inconsistent, leading to terrible forecast accuracy.

This is the dark side of the AI and sales tech revolution. Every tool promises to make you more efficient. But when those tools operate in silos, they collectively make you less efficient. You're drowning in capabilities but starving for integration.

From experience, this is what the Unify stage of the QUANTUM Framework addresses. You can qualify prospects brilliantly, understand them deeply, anticipate their needs, steer through complex deals, tailor your approach, and measure everything. But if all these activities happen in disconnected tools with information trapped in silos, you're working five times harder than necessary.

Unify is about creating an integrated sales ecosystem where intelligence flows smoothly, insights compound across systems, and AI can orchestrate activities across your entire tech stack. It's about moving from a collection of point solutions to a unified operating system for AI-powered sales.

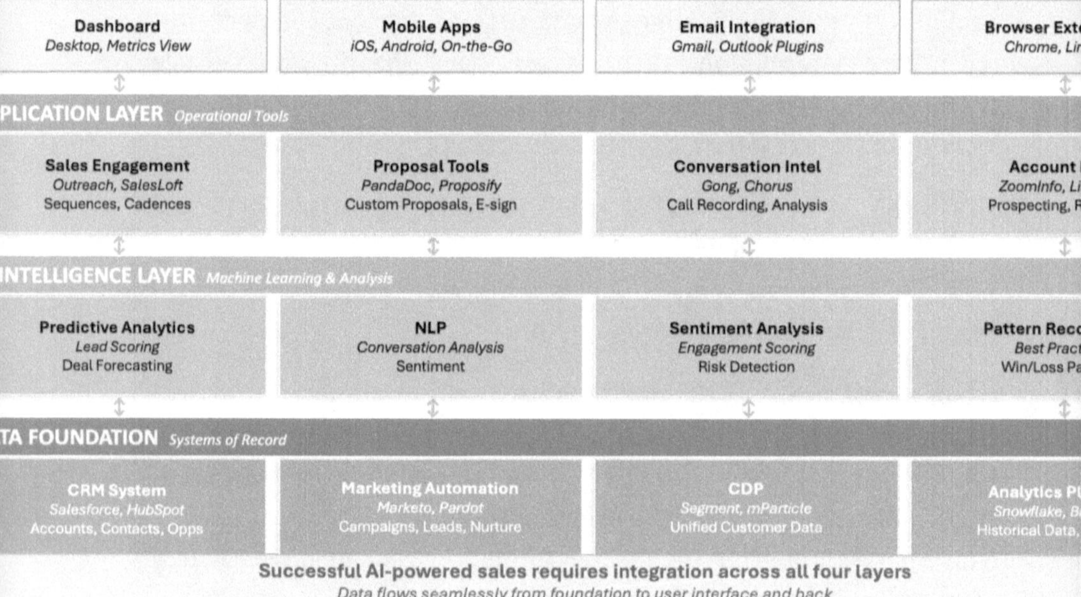

Figure 8.1: Four-Layer Integrated AI Sales Technology Stack

The Integration Crisis

Let's start with an uncomfortable truth: most sales organisations have an integration problem they don't fully recognise.

Ask any sales leader if they have good systems, and they'll probably say yes. They've invested in CRM, sales engagement platforms, conversation intelligence, and various AI tools. Each purchase made sense individually. But ask that same leader if their systems work together effortlessly, and you'll get a different answer.

Here's what typically happens:

The data doesn't flow. Information entered in one system doesn't automatically appear in others. Sales reps manually copy prospect details from LinkedIn to Salesforce to Outreach. Notes from calls don't automatically update opportunity records. Email interactions tracked in one place don't sync with activity logs in another.

The insights don't connect. Your conversation intelligence tool identifies that prospects from financial services always ask about compliance. But that insight doesn't automatically flow to your sales engagement system to adjust messaging for financial services prospects. It just sits there, available if someone remembers to look, but not actually improving how the team operates.

From a practical standpoint, the workflows break at boundaries. You've automated prospecting sequences in your sales engagement platform. But when a prospect responds and expresses interest, someone must manually create an opportunity in your CRM, schedule a meeting, and trigger the next workflow. The handoff points between systems require human intervention.

The reporting is fragmented. Your CRM shows pipeline metrics. Your conversation intelligence platform shows call analytics. Your sales engagement tool shows email performance. But getting a unified view of what's working—which activities drive which outcomes—requires exporting data from multiple sources, combining it manually, and building custom reports.

This isn't just inefficient. It's actively harmful. Every manual data entry creates opportunity for error. Every system boundary that requires human intervention creates delays. Every fragmented view of performance makes it harder to optimise. And every hour your sales team spends on administrative work is an hour they're not actually selling.

The promise of AI-powered sales is that technology handles the tedious work so humans can focus on judgment and relationships. But if your systems aren't unified, you're getting the opposite: technology is creating more work, not less.

What Unification Actually Means

Unification isn't just about connecting tools through APIs. It's about creating an ecosystem where:

Data flows automatically. Information entered once propagates to all relevant systems without manual intervention. When a prospect's job title changes in LinkedIn Sales Navigator, it updates in your CRM, triggers adjustments to your messaging in your sales engagement platform, and informs the AI about how to adjust its predictions.

Insights compound across systems. When your conversation intelligence tool identifies a pattern in customer calls, that insight automatically adjusts targeting in your prospecting tool, refines messaging in your email sequences, updates risk predictions in your deal forecasting system, and informs content recommendations in your proposal software.

Workflows orchestrate cleanly. AI can trigger multi-step workflows that span multiple systems without requiring human intervention at each handoff. A prospect's behaviour on your website can automatically create a lead, add them to the appropriate cadence, personalise the messaging based on their profile, schedule follow-up tasks, and adjust based on their engagement. All without a human touching it.

Intelligence is centralized. All your systems feed into and draw from a single source of intelligence. Every interaction, every data point, every insight contributes to a unified understanding of each prospect and overall patterns. The AI isn't learning separately in each tool. It's learning end-to-end across your entire operation.

When systems are truly unified, the whole becomes dramatically greater than the sum of its parts. Each tool amplifies the value of the others. The insights from conversation intelligence make your prospecting more targeted. Better prospecting data makes your email sequences more relevant. More relevant sequences generate better conversations. Better conversations provide richer intelligence. The cycle reinforces itself.

The Four Layers of Sales Technology

Before we dive into how to unify systems, it helps to understand how sales technology stacks are typically structured. Most evolved sales operations have four distinct layers:

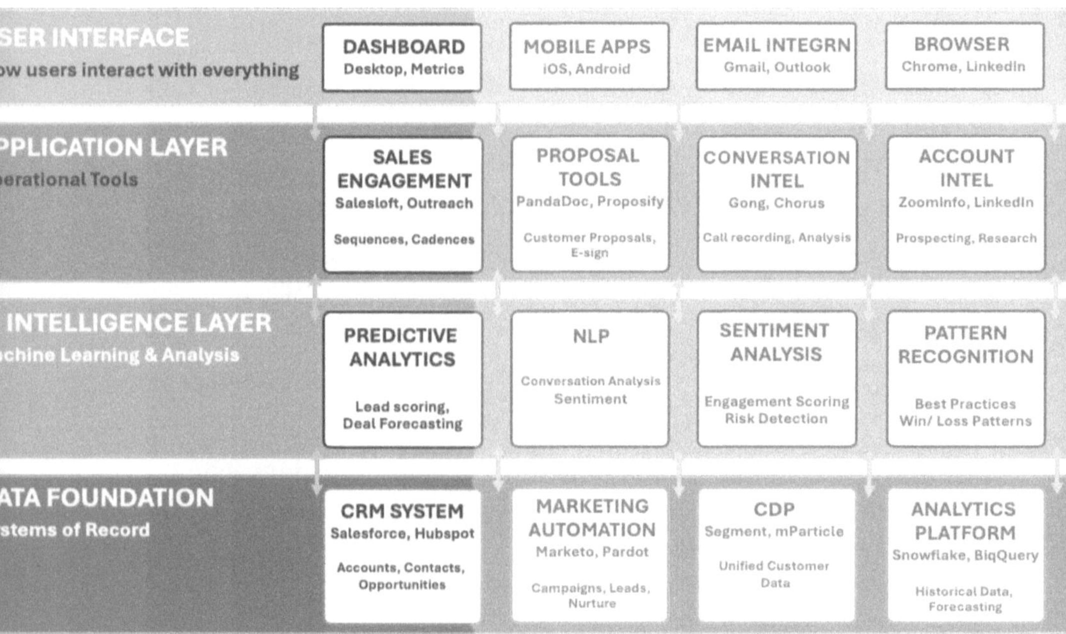

Figure 8.2: Four-Layer Integrated AI Sales Technology Stack. Four-Layer Architecture for Unified Sales Intelligence

This is where information lives. Your systems of record that store customer data, interaction history, and transactional information.

Core components:

- **CRM (Salesforce, HubSpot, Microsoft Dynamics):** Stores accounts, contacts, opportunities, and activities
- **Marketing automation (Marketo, Eloqua, Pardot):** Tracks marketing interactions and nurture campaigns
- **Customer data platform (Segment, particle):** Unifies customer data from multiple sources
- **Data warehouse (Snowflake, Big Query):** Stores historical data for analysis

Across industries, the data foundation is critical because it's the source of truth. If this layer is fragmented. Customer information stored differently across systems, inconsistent definitions of what constitutes a "qualified lead," conflicting records for the same person. Everything built on top of it becomes unreliable.

Notice how this is where machine learning and AI process data to generate insights, predictions, and recommendations.

Core components:

- **Predictive analytics engines:** Forecast deal outcomes, score leads, identify churn risk
- **Natural language processing:** Analyse sales calls, emails, and proposals to extract insights
- **Recommendation engines:** Suggest next best actions, content to share, and stakeholders to engage
- **Pattern recognition systems:** Identify what works across deals and surface best practices

The intelligence layer is what makes modern sales technology "smart." But it's only as good as the data it can access. If AI tools can only see fragments of the picture. Calls but not emails, emails but not CRM activity, CRM activity but not conversation content. Their predictions and recommendations are based on incomplete information.

Layer 3: Application Layer

This is where work happens. The tools sales reps use daily to execute sales activities.

Core components:

- **Sales engagement platforms (Outreach, SalesLoft):** Automate outbound sequences and track engagement
- **Conversation intelligence (Gong, Chorus.ai):** Record and analyse sales calls
- **Proposal and quote tools (Panadol, Conga):** Generate and track proposals
- **Meeting scheduling (Calendly, Chili Piper):** Coordinate calendars and book meetings
- **Content management (High spot, Seismic):** Organise and recommend sales collateral

These are the most visible tools because salespeople interact with them constantly. But their effectiveness depends entirely on whether they're connected to the intelligence and data layers below them.

Layer 4: User Interface

Notice how this is how salespeople access information and act. The dashboards, mobile apps, and notifications that surface insights when they're needed.

Core components:

- **CRM interface:** Where reps view accounts and manage opportunities

- **Mobile applications:** Access critical information from phones and tablets
- **Email integration:** Work within existing email workflow rather than separate tools
- **Browser extensions:** Surface relevant information without leaving current context
- **Notification systems:** Alert reps to important events and recommended actions

The user interface layer is often overlooked in discussions of sales technology, but it's crucial. If AI generates brilliant insights but surfaces them in a separate dashboard that reps never check, those insights might as well not exist. Unification requires bringing intelligence to where people are already working, not asking them to go somewhere new to find it.

The Integration Architecture

Now that we understand the layers, let's talk about how to unify them. There are three fundamental integration patterns:

Pattern 1: Hub-and-Spoke (CRM-Centric)

In this model, your CRM serves as the central hub, and all other tools connect to it through integrations. Data flows into and out of the CRM, which acts as the system of record.

Advantages:

- Simple conceptual model. Everything revolves around one system
- Most tools have pre-built CRM integrations
- Sales reps primarily work in CRM, so data is where they need it

Disadvantages:

- CRM becomes a bottleneck. If it goes down or has limitations, everything is affected

- Not all tools integrate directly with CRM, requiring custom development
- Data often flows one direction (into CRM) but not the other direction efficiently

Best for: Organisations with established CRM implementations where the sales process is already CRM-centric and most tools have native integrations.

Pattern 2: iPaaS (Integration Platform as a Service)

In this model, you use a middleware integration platform (like Zapier, Workmate, or MuleSoft) that sits between your tools and orchestrates data flow according to rules you define.

Advantages:
- Flexible—can connect any tools together, not just to CRM
- Easier to build complex workflows that span multiple systems
- Can route data intelligently based on conditions and business logic
- Doesn't overburden any single system as the integration point

Disadvantages:
- Another system to manage and pay for
- Can become complex with many integrations and rules
- Creates dependency on the iPaaS platform itself

Best for: Organisations with diverse tool stacks where different teams use different primary systems, or where complex workflows require sophisticated orchestration.

Pattern 3: Data Warehouse / Reverse ETL

In this modern approach, all tools feed data into a central data warehouse (like Snowflake or Big Query), where it's cleaned, unified, and enriched. Then "reverse ETL" tools (like Census or High touch) push the unified data back out to operational systems.

Advantages:

- Complete historical data in one place for advanced analytics
- Can unify data from sources that don't directly integrate
- Create single source of truth for key entities (contacts, accounts, opportunities)
- Enables sophisticated AI/ML that requires complete datasets

Disadvantages:

- Requires data engineering expertise to implement and maintain
- Initial setup is complex and time-consuming
- Real-time synchronisation can be challenging
- Higher technical requirements than other approaches

Best for: Larger organisations with dedicated data teams, or companies where advanced analytics and AI are critical competitive advantages requiring complete data access.

Most successful sales organisations end up using a hybrid approach. CRM as the operational hub for daily sales work, iPaaS for workflow automation and tactical integrations, and data warehouse for analytics and AI/ML that require complete historical data.

Building Your Unified Sales Stack

Let's get tactical. Here's how to build a unified sales technology ecosystem:

Step 1: Audit Your Current State

Start by mapping what you have. For each tool:

Document the tool:

- What does it do?
- Who uses it?
- What data does it contain?
- How does it integrate (or not) with other systems?

Identify the gaps:

- What data should flow between systems but doesn't?
- What manual work could be automated if systems were connected?
- What insights are trapped in one system that would be valuable elsewhere?
- What duplicate data entry happens because systems don't sync?

Quantify the cost:

- How much time do reps spend on manual data entry and synchronisation?
- How often do opportunities get missed because information doesn't flow?
- What errors occur because data is inconsistent across systems?
- How much revenue is at risk due to poor visibility caused by fragmentation?

This audit typically reveals that integration problems are costing far more than expected. Both in direct time waste and in missed opportunities.

Step 2: Define Your Integration Priorities

You can't unify everything at once. Prioritise based on impact:

High-priority integrations (do these first):

- CRM ↔ Sales engagement platform: Ensure outbound activity automatically logs to CRM and CRM data flows to engagement tool
- Conversation intelligence → CRM: Push call insights and action items automatically to opportunity records
- Lead sources → CRM: Automate lead creation from web forms, events, and other sources

- CRM → Analytics/BI: Enable reliable reporting on pipeline and activities

Medium-priority integrations (do next):

- Prospecting data (ZoomInfo, etc.) → CRM and sales engagement: Enrich records and trigger sequences automatically
- Proposal tools ↔ CRM: Track proposal status and automatically update opportunities when proposals are sent/viewed/signed
- Marketing automation ↔ CRM: Sync lead scoring and track marketing touchpoints in sales context
- Calendar/scheduling tools → CRM: Log meetings automatically

Lower-priority integrations (nice to have):

- Content management → Various tools: Surface relevant content recommendations
- Social selling tools → CRM: Track social interactions
- Contract management → CRM: Sync contract status
- Customer success tools ↔ CRM: Bridge sales-to-post-sale handoff

Looking closer, the goal is to start with integrations that eliminate the most painful manual work and open up the most valuable insights, then progressively expand the unified ecosystem.

Step 3: Choose Your Integration Approach

Based on your technical resources and requirements:

If you have limited technical resources:
- Use tools with native integrations to each other (e.g., Salesforce + Outreach + Gong have tight integrations)
- Use no-code iPaaS tools like Zapier for simple workflows
- Prioritise tools that market themselves as "open" and integration-friendly

- Accept some limitations in customisation to gain ease of implementation

If you have moderate technical resources:

- Use iPaaS platform like Workmate or Tray.io for sophisticated workflow orchestration
- Build custom integrations for unique workflows that pre-built connectors don't support
- Invest in data quality and governance so integrated data is useful
- Create feedback loops where users report integration issues and you continuously improve

If you have strong technical resources:

- Consider data warehouse approach with reverse ETL for maximum flexibility
- Build custom integrations and middleware where commercial options have limitations
- Invest in real-time data pipelines for immediate insight propagation
- Create unified data models that ensure consistency across systems

Most organisations start with the first approach and graduate to more sophisticated integration as both their technical capabilities and requirements grow.

Step 4: Implement Data Governance

Integration without governance creates unified chaos instead of unified clarity. Establish:

Data standards:

- How are accounts and contacts matched across systems? (By email? Company domain? Salesforce ID?)

- What fields are required? What are the acceptable values?
- Who is the source of truth for which data elements?
- How do you handle conflicts when the same data exists differently in multiple systems?

Ownership and accountability:

- Who is responsible for data quality in each system?
- Who troubleshoots when integration breaks?
- Who decides what new integrations get built?
- Who maintains documentation of how systems connect?

Change management:

- How do you test integration changes before deploying to production?
- How do you communicate when integration behaviour changes?
- How do you train users on new unified workflows?
- How do you measure whether integration is improving productivity?

The technical integration is often easier than the organisational change management. Systems can be connected in days; getting people to use those connections effectively can take months.

Step 5: Build Intelligence on Top of Integration

Once data flows naturally, you can build AI capabilities that apply the complete picture:

Cross-system insights:

- Correlate which email messages (from sales engagement platform) lead to successful calls (from conversation intelligence) that lead to closed deals (from CRM)
- Identify which combination of touchpoints and stakeholders predicts deal success

- Surface patterns across the entire sales cycle, not just fragments visible in individual tools

Intelligent orchestration:

- Use AI to determine which action to take next based on complete context. Should you send an email, make a call, involve a different stakeholder, or wait?
- Automatically route leads to the right rep based on full context of skills, current workload, and success patterns
- Adjust sales engagement sequences in real-time based on conversation intelligence insights

Predictive capabilities:

- Forecast deal outcomes using data from CRM, conversation analysis, stakeholder engagement, and historical patterns
- Predict which prospects are most likely to engage based on their complete profile and behaviour
- Identify deals at risk by analysing patterns across all touchpoints, not just CRM data

This is where unification pays off exponentially. Individual tools provide linear improvements. Unified systems with AI orchestration provide exponential improvements because insights compound across the ecosystem.

Common Integration Pitfalls

Let me save you from the mistakes I've seen repeatedly:

Pitfall 1: Over-Engineering the Solution

I've watched companies spend months building complex integration architectures when simple pre-built connectors would have worked fine. Start simple. Connect your two or three most critical systems with out-of-the-box integrations, prove value, then expand.

Don't let perfect be the enemy of good. A basic integration that works today is infinitely better than a sophisticated architecture that will be ready in six months.

Pitfall 2: Integrating Bad Data

Connecting systems doesn't fix data quality. It amplifies it. If your CRM has duplicate records, incomplete information, and inconsistent formatting, integrating it with other tools just propagates that mess everywhere.

Clean your data before integrating. Or at minimum, build data quality checks into your integration workflows so bad data gets flagged rather than spread.

Pitfall 3: Ignoring the User Experience

I've seen technically perfect integrations that users actively worked around because the experience was terrible. Data synced flawlessly, but the notifications were annoying, the workflows interrupted natural selling, and the overhead of using the integrated system was worse than the manual process.

Design integration from the user perspective. How will this make their work easier? What do they need to see and when? How can you surface insights without overwhelming them? Technical success without user adoption is just expensive failure.

Pitfall 4: Building Dependencies on Fragile Systems

When you integrate systems, you create dependencies. If one tool goes down or has issues, it can cascade across your entire stack. I've watched sales teams unable to work because their sales engagement platform had an outage and all their workflows depended on it.

Build resilience into your integrations:

- Have fallback workflows when systems are unavailable

- Don't block critical actions on integration success. Allow async updates
- Monitor integration health and have alerts when things break
- Document manual workarounds for when automation fails

Pitfall 5: Set It and Forget It

Integration isn't a one-time project. It's ongoing maintenance. Tools change their APIs. Business requirements evolve. Data models shift. What worked perfectly six months ago might not work today.

Assign ownership for integration health. Monitor it continuously. Budget time for maintenance and improvements. Treat your integrated sales stack as critical infrastructure that requires ongoing investment, not a completed project.

The Unified Sales Workflow: A Day in the Life

Let me paint a picture of what unified systems look like in practice. Here's how a sales rep's day changes:

8:00 AM - Morning Prioritisation

Before unification: Rep logs into five different tools to figure out what to do today. Checks CRM for scheduled calls. Reviews Outreach to see which prospects replied. Opens Gong to see if any calls need follow-up. Looks at LinkedIn for new activity. Manually builds a to-do list.

After unification: Rep opens a single dashboard that AI has already prioritised based on complete context:

- "Top priority: Follow up with Acme Corp. Your champion mentioned budget approval in yesterday's call (Gong detected positive signal), and they opened your proposal three times this morning (tracked by Docent)"

- "Second priority: Call TechStars Inc. They're in your ICP, engaged with your last two emails, and their CFO just changed (LinkedIn signal suggests buying window)"
- AI has already drafted follow-up emails using insights from recent calls and adjusted the messaging based on what's worked with similar prospects

10:00 AM - Discovery Call

Before unification: Rep conducts call, takes notes manually, remembers to log the call in CRM afterward, updates opportunity stage, and hopes they didn't miss any important details in their notes.

After unification: Call is automatically recorded by conversation intelligence. AI transcribes it, identifies key points (budget mentioned: $500K, timeline: Q2, decision-makers: CFO and CTO), extracts action items (send security documentation, schedule technical demo), and automatically:

- Updates the opportunity in CRM with new budget and timeline information
- Creates follow-up tasks with due dates based on commitments made
- Adjusts the opportunity score based on positive buying signals detected
- Triggers personalised email sequence with the specific security documentation mentioned
- Alerts rep that similar prospects needed technical validation from IT, suggesting proactive stakeholder expansion

2:00 PM - Proposal Creation

Before unification: Rep manually creates proposal by copying details from CRM, email threads, and call notes. Uses last quarter's proposal as template, manually adjusts pricing, hopes all information is accurate and current.

After unification: Proposal tool pulls complete opportunity data from CRM, including:

- Prospect's specific pain points mentioned across all calls (from conversation intelligence)
- Relevant case studies from similar companies (matched by AI based on complete prospect profile)
- Accurate pricing based on actual requirements discussed (captured and validated across systems)
- Personalised value proposition using language that's resonated in previous conversations AI suggests adding a section on integration capabilities because this prospect's tech stack (enriched from Built With data integrated into CRM) indicates they heavily use Salesforce, and your integration story tests well with Salesforce shops.

4:00 PM - Deal Review

Before unification: Sales manager asks rep for update. Rep provides verbal overview based on what they remember, plus whatever is documented in CRM (which is usually incomplete).

After unification: Manager sees complete deal health dashboard:

- Engagement trending down over past two weeks (email open rates declining, champion taking longer to respond)
- Conversation intelligence flagged that champion mentioned "budget review" in last call. Possible risk signal
- No engagement with economic buyer (CFO) yet, which historically predicts stalled deals at this stage
- Competitive threat: prospect visited competitor's website (tracked via reverse IP lookup integrated into the stack) Manager and rep discuss specific interventions: involve CFO earlier, address budget concerns proactively, differentiate from competitor.

6:00 PM - End of Day

Before unification: Rep spends 30 minutes updating CRM with the day's activities, copying information from various tools, making sure everything is logged for the forecast meeting tomorrow.

After unification: All activities already synced automatically. Rep spends five minutes reviewing AI-generated summary of the day and tomorrow's recommended priorities. Everything is already documented, forecasts are automatically updated based on actual signals, and the system has prepared tomorrow's workflow.

Time saved: 2+ hours per day. **Quality improvement:** Massive reduction in missed follow-ups, faster response times, more accurate forecasting, better-informed decisions.

That's the promise of unification. Not just connecting tools but fundamentally changing how selling works by eliminating administrative overhead and amplifying intelligence.

Measuring Unification Success

How do you know if your integration efforts are working? Track these metrics:

Efficiency metrics:

- Time spent on administrative work (should decrease significantly)
- Data entry errors and duplicates (should decline)
- Time from lead creation to first outreach (should shorten)
- Average time to update opportunity after important events (should approach zero with automation)

Adoption metrics:

- Percentage of reps actively using integrated workflows
- Volume of data flowing between systems (should be high and consistent)

- Number of manual workarounds being used (should decrease over time)

Outcome metrics:

- Forecast accuracy (should improve with better data quality)
- Sales cycle length (should decrease with better coordination)
- Win rate (should improve with better intelligence and timing)
- Revenue per rep (should increase as administrative burden decreases)

The goal isn't integration for its own sake. It's measurable business impact. If you're connecting systems but not seeing improvements in these metrics, something isn't working.

The Future of Unified Sales

We're still in early days of truly unified sales ecosystems. Most organisations have achieved basic integration. Data syncing between a few core systems. But we're rapidly moving toward something more sophisticated:

Autonomous orchestration: AI that doesn't just recommend actions but takes them across multiple systems based on complete context. When a prospect shows buying signals, AI automatically adjusts sequences, creates opportunities, schedules meetings, and prepares personalised materials. Without human intervention.

Real-time intelligence: Systems that don't just batch-sync overnight but update instantly across the entire stack. When something important happens in any system, all other systems know immediately and adjust accordingly.

Natural language interfaces: Instead of logging into multiple tools, sales reps simply tell AI what they need, "Show me my highest-priority opportunities" or "What should I focus on with Acme Corp today?". And AI pulls complete context from across the unified stack and surfaces exactly what's needed.

Self-healing systems: Integration that monitors itself, detects when something breaks, and either fixes it automatically or alerts the right people with specific diagnostics.

Put simply, this isn't science fiction. Companies like Salesforce (Einstein), HubSpot (HubSpot AI), and newer players like Clari are already building toward this vision. The question isn't whether this future is coming, but whether your sales organisation will be positioned to take advantage of it.

Getting Started: Your 90-Day Unification Plan

If you're reading this and realising your systems are fragmented, here's a practical 90-day plan to start unifying:

Days 1-30: Audit and Prioritise

- Map your current tech stack and integration state
- Identify the three most painful manual workflows
- Quantify time wasted and opportunities missed due to fragmentation
- Get executive sponsorship by showing cost of status quo

Days 31-60: Quick Wins

- Implement your three highest-priority integrations using pre-built connectors
- Focus on basic data flow (CRM ↔ sales engagement, conversation intelligence → CRM)
- Train reps on new unified workflows
- Measure time saved and adoption rates

Days 61-90: Expand and Optimise

- Add next tier of integrations based on learnings from first phase
- Start building AI capabilities on top of unified data

- Establish data governance and quality processes
- Create roadmap for ongoing integration improvements

Don't try to unify everything immediately. Start with core workflows, prove value, gain momentum, then expand systematically.

The Unification Imperative

Here's the bottom line: you cannot achieve AI-powered sales excellence with fragmented systems. You can have the best AI tools, the smartest strategies, and the most talented reps, but if information is trapped in silos and insights don't flow naturally, you're operating at a fraction of your potential.

Unification isn't a nice-to-have. It's the foundation that makes everything else in the QUANTUM Framework work. Qualification, understanding, anticipation, navigation, and tailoring all depend on having complete, accessible, actionable intelligence. Measurement requires unified data to identify what drives results.

Critically, the companies that dominate sales in the next decade won't just be those with the best individual tools. They'll be the ones who've built truly integrated sales ecosystems where human intelligence and artificial intelligence work together smoothly across every interaction.

In the next chapter, we'll tackle Measure. How to track what matters, identify leading indicators of success, and continuously optimise your AI-powered sales operation. Because unified systems are only as valuable as your ability to learn from what they're telling you.

CHAPTER OUTCOME

- [ ] An inventory of your current tools and what each is actually used for
- [ ] A list of your top data problems (silos, duplication, missing fields, inconsistent definitions)

☐ A "minimum viable integration" plan (what must connect first to create value)

☐ A chosen integration approach (hub-and-spoke, iPaaS, warehouse, etc.) and why

☐ A governance rule-set: who owns data quality, definitions, access, and change control

☐ A plan to reduce workflow friction (simplify UX, remove duplicate entry, automate handoffs)

☐ A staged rollout plan (foundation → expansion → optimisation)

CHAPTER 9

Measure – Analytics and Optimisation

By the end of this chapter, you will be able to:

- Separate leading vs lagging indicators and define the few metrics that actually predict outcomes.

- Use AI-enhanced analytics (real-time scoring, cross-deal pattern recognition, forecasting, anomaly detection).

- Build a measurement system end-to-end (metrics → baselines → dashboards → cadence → feedback loop).

- Instrument performance at rep level and generate coaching insights without drowning in noise.

- Avoid common measurement failures (vanity metrics, lagging-only views, analysis paralysis, no action).

Tracking what matters and continuously improving performance

Three years into running a sales team at a fast-growing SaaS company, I thought I had measurement figured out. We tracked everything: calls made, emails sent, meetings booked, pipeline created opportunities in each stage, win rates, average deal size, sales cycle length. Our weekly forecast meetings involved reviewing spreadsheets with dozens of metrics.

Then one quarter, despite hitting all our activity metrics—more calls than ever, higher meeting rates, strong pipeline generation—we missed revenue by 30%. Every leading indicator looked healthy, yet the lagging indicator that mattered (closed revenue) told a different story.

I realised we were measuring extensively but not measuring effectively. We were tracking what was easy to count rather than what predicted success. Our metrics were historical (what already happened) rather than predictive (what's about to happen). We were measuring activity rather than effectiveness.

That quarter forced a complete rethinking of our measurement approach. We stopped celebrating activity volume and started focusing on quality indicators. We stopped looking backward at closed deals and started looking forward at deal health signals. We stopped measuring individuals in isolation and started measuring the effectiveness of our entire sales system.

This is what the Measure stage of the QUANTUM Framework addresses. You've qualified the right prospects, understood them deeply, anticipated their needs, navigated complexity, tailored your approach, and unified your systems. Now you need to measure what's working so you can continuously optimise and improve.

The goal isn't measurement for measurement's sake. It's building a feedback loop that makes your entire sales operation progressively more effective over time.

The Measurement Problem

Most sales organisations suffer from one of two measurement problems:

Problem 1: Measuring Too Little

Some teams barely measure anything beyond closed revenue. They know their monthly and quarterly numbers but have no insight into what's driving them. When results are bad, they can't diagnose why. When results are good, they don't know which behaviours to reinforce. They're flying blind.

Problem 2: Measuring Too Much

Other teams measure everything and drown in data. They have dashboards with 47 metrics, weekly reports that take hours to compile, and analysts who spend more time pulling numbers than anyone spends acting on them. They have data but no insights, information but no understanding.

The solution is strategic measurement: tracking the specific metrics that predict outcomes and drive decisions, while ignoring vanity metrics that just consume time without providing actionable intelligence.

Leading vs. Lagging Indicators

In practice, the foundation of effective measurement is understanding the difference between leading and lagging indicators:

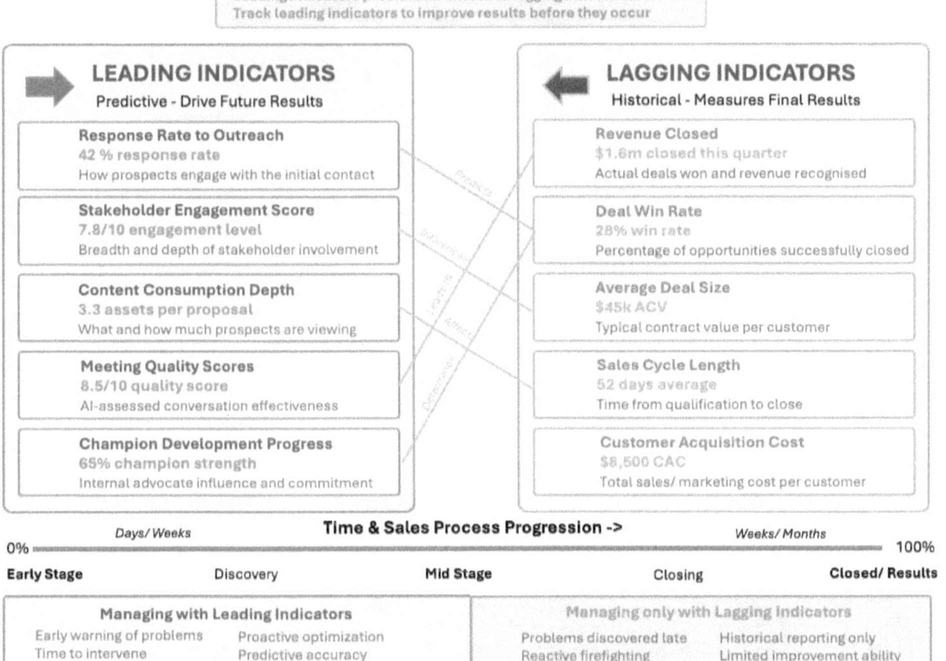

Figure 9.1: Leading vs. Lagging Indicators. Predicting vs. Measuring Sales Performance – Predict Future Performance vs. Measure Past Results

Lagging indicators tell you what already happened. They're outcomes. Revenue closed, deals won, customers acquired. These are important because they're what you're ultimately measured on, but they're historical. By the time you see problems in lagging indicators, it's too late to fix them for that period.

Leading indicators predict what's about to happen. They're the early signals that forecast future outcomes. Prospect engagement, deal progression velocity, pipeline quality. When you see problems in leading indicators, you have time to intervene before they become revenue misses.

Elite sales organisations obsess over leading indicators because they provide early warning and opportunity for correction. They track lagging indicators to measure results, but they manage the business using leading indicators.

Critical Leading Indicators

Here are the leading indicators that consistently predict sales success:

Pipeline Quality Score: Not just pipeline dollar value, but quality-adjusted pipeline that accounts for deal probability, age, and engagement level. A $5M pipeline of low-quality deals is worse than a $3M pipeline of high-quality deals.

Stakeholder Engagement Velocity: How quickly are you expanding stakeholder engagement in your opportunities? Deals that stay single-threaded stall. Deals where engagement expands to multiple stakeholder's progress.

Response Time and Responsiveness: How quickly do prospects respond to your outreach? How often do they accept meetings versus decline? Declining responsiveness predicts stalled deals before they officially stall.

Champion Strength: How actively is your champion advocating internally? Conversation intelligence can measure this by tracking champion language in calls. Do they use "we" language suggesting they're part of the solution, or "you/they" language suggesting they're just evaluating?

Content Engagement Depth: Which content are prospects consuming, and how deeply? If they're reading case studies and pricing pages, that's stronger intent than just downloading a generic whitepaper.

Deal Progression Velocity: How quickly are deals moving through stages? Deals that progress steadily close. Deals that sit in one stage for weeks typically die.

Technical Validation Status: In complex sales, has technical due diligence been completed successfully? Technical blockers kill deals late in the cycle. Early technical validation is a strong leading indicator.

Budget Confirmation Specificity: Not just "do they have budget?" but "has budget been formally allocated, and do we know the specific amount?" Vague budget discussions predict slippage.

Critical Lagging Indicators

You still need to track outcomes to measure ultimate success:

Revenue Attainment: Actual closed revenue versus quota, by rep and team.

Win Rate: Percentage of opportunities that close successfully, overall and by segment.

Average Deal Size: Typical contract value, tracked over time to identify trends.

Sales Cycle Length: Average time from opportunity creation to close, by deal type.

Customer Acquisition Cost (CAC): Total sales and marketing spend divided by new customers acquired.

Pipeline Coverage: Ratio of pipeline to quota (typically 3-4x for healthy coverage).

The key is using lagging indicators to evaluate results while using leading indicators to drive daily/weekly decisions and interventions.

AI-Enhanced Analytics

Traditional sales analytics rely on manual data entry, human analysis, and retrospective reporting. By the time you spot problems, momentum is lost. AI changes this completely by enabling:

Real-Time Deal Scoring

AI can analyse every deal in your pipeline continuously and score them based on probability to close:

"This deal has strong champion engagement and multiple stakeholders involved (positive signals), but timeline has slipped twice and economic buyer hasn't engaged yet (negative signals). Current close probability: 42%, down from 58% last week. Recommend intervention."

Instead of relying on rep gut feel or stage-based forecasting, you get data-driven predictions that update in real-time as new information emerges.

Pattern Recognition Across Deals

AI can identify patterns humans would never spot:

"Deals where technical validation happens in week 3-4 close at 67% rate. Deals where it happens after week 6 close at 23% rate. Recommend moving technical discussions earlier in cycle."

"When CFO joins calls before week 5, deals close 2.1x faster. When CFO only engages late in cycle, average sales cycle extends by 40 days."

These insights let you optimise your process based on what predicts success.

Predictive Forecasting

Instead of asking reps, "what's going to close this quarter?" and hoping they're right, AI analyses deal characteristics and behaviours to predict outcomes:

"Based on current pipeline health, engagement patterns, and historical conversion rates, forecast for quarter is $2.8M-$3.2M with 80% confidence. This is below $3.5M target. Recommend accelerating 5 specific deals identified as most likely to pull forward."

Taken together, this gives you accurate forecasts and specific guidance on where to focus.

Automated Anomaly Detection

AI can flag unusual patterns that warrant investigation:

"Deal with Acme Corp has been in proposal stage for 45 days with no stakeholder engagement in last 2 weeks. This is 3x typical time in stage and engagement drop-off exceeds threshold. Recommend immediate re-qualification."

Instead of waiting for weekly pipeline reviews to spot problems, issues are flagged immediately when patterns diverge from healthy norms.

Building Your Measurement System

Here's how to implement effective measurement systematically:

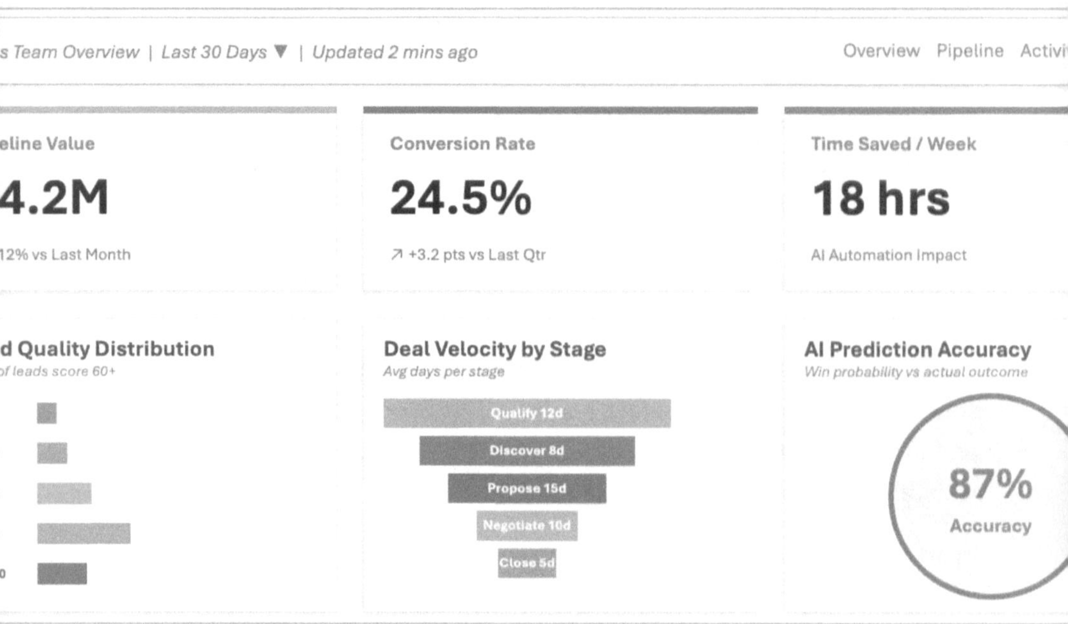

Figure 9.2: AI-Powered Sales Analytics KPI Dashboard.
Real-Time Performance Monitoring

Don't track everything. Identify the 8-12 metrics that drive decisions:

Pipeline Metrics (3-4):

- Qualified pipeline created
- Pipeline quality score
- Pipeline coverage ratio
- Average deal size

Activity Effectiveness Metrics (2-3):

- Qualified opportunity conversion rate
- Meeting-to-opportunity rate
- Demo-to-proposal rate

Deal Health Metrics (2-3):

- Average deal score/probability
- Multi-threading percentage (deals with 3+ stakeholders engaged)
- Deal velocity (days per stage)

Outcome Metrics (2-3):

- Win rate
- Revenue attainment
- Sales cycle length

Step 2: Establish Baselines and Targets

For each metric, know:

- **Current performance:** Where are we now?
- **Historical trends:** How has this changed over time?
- **Target performance:** Where do we need to be?
- **Best-in-class benchmark:** What's excellent performance look like?

Example: "Current win rate is 22%. Over past 6 months, it's been declining from 28%. Target is 25%. Top quartile of reps achieve 35%+."

This context turns data into intelligence. You know whether you're improving or declining, whether you're on track or behind, and how much upside exists.

Step 3: Create Visual Dashboards

Metrics are only useful if people see them and act on them. Build dashboards that:

Show trends, not just snapshots: Don't just show "win rate is 22%". Show win rate trend over past 6 months so you can see direction.

Highlight exceptions: Use colour coding to flag metrics that are below target (red), at risk (yellow), or exceeding expectations (green).

Provide drill-down capability: If win rate is declining, can you drill down to see which segment or rep or deal type is driving the decline?

Update in real-time: Static weekly reports are too slow. Dashboards should reflect current data so you can respond quickly.

Are accessible to those who need them: Sales reps see their individual metrics. Managers see team metrics. Executives see organisational metrics. Everyone gets relevant data without being overwhelmed.

Step 4: Establish Cadence for Review

Don't just collect data. Create rhythms for using it:

Daily: Reps review their priority deals and deal health scores, acting on deals flagged as at-risk.

Weekly: Team reviews pipeline health, win rates, and activity effectiveness. Identify issues and course-correct quickly.

Monthly: Leadership reviews trends, identifies systemic issues, adjusts strategy and process based on what's working and what's not.

Quarterly: Deep analysis of win/loss patterns, sales cycle optimisation, conversion funnel, market trends. Strategic decisions about process changes, training focus, tool investments.

Step 5: Close the Feedback Loop

The point of measurement is optimisation. When you identify what's working, do more of it. When you identify what's not working, fix it.

Example feedback loop:

Measure: AI flags that deals with early CFO engagement close 2x faster.

Insight: Economic buyer engagement earlier in cycle is a critical success factor.

Action: Adjust sales process to mandate CFO conversation by week 4 of every qualified opportunity.

Result: Sales cycle length decreases by 18% over next quarter.

Optimisation: Continue reinforcing this behaviour, track to ensure it's sustained.

This is how measurement drives continuous improvement. Not just reporting what happened but systematically getting better.

Measuring What Matters to Individuals

Beyond organisational metrics, effective measurement gives each rep visibility into their own performance and clear guidance on improvement:

Rep-Level Dashboards

Each rep should see:

Pipeline health: Quality-adjusted pipeline versus their quota (are they on track?)

Deal prioritisation: Which deals need attention today based on health scores and engagement patterns

Activity effectiveness: Are their calls, emails, and demos converting at expected rates?

Skill development: Where are they strong versus where do they need coaching? (Demo skills, objection handling, multi-threading, etc.)

Comparison to benchmarks: How do they compare to team average and top performers? (Not to shame, but to identify improvement opportunities)

Coaching Insights from AI

Conversation intelligence provides coaching gold for managers:

"Rep is doing well at discovery—asking good questions, listening actively. Weak area is economic buyer engagement—in 12 calls this month, only mentioned budget/authority twice. Recommend coaching on qualification questioning."

"Rep's win rate is 15% below team average. Analysis shows deals are lost late in cycle after technical validation. Issue appears to be inadequate technical discovery early on. Recommend shadowing senior technical rep on next 3 technical calls."

Notice how this moves coaching from generic advice ("be better at discovery") to specific, data-driven guidance ("you need to engage economic buyers earlier. Here's evidence and here's how").

Measuring AI Effectiveness

If you're using AI throughout your sales process (you should be), you need to measure whether it's helping:

AI Impact Metrics

Time savings: How much time does AI save on research, content generation, data entry, reporting? Track actual hours saved per rep per week.

Quality improvement: Do AI-assisted emails get better response rates? Do AI-generated proposals have higher win rates?

Deal health improvement: Do deals where AI recommendations are followed perform better than those where they're ignored?

Forecast accuracy: How much has predictive AI improved forecast accuracy versus human judgment alone?

Adoption rate: What percentage of reps are using AI tools effectively versus ignoring them?

If AI isn't demonstrably improving outcomes, either you're measuring wrong, using it wrong, or it's the wrong tool.

Common Measurement Mistakes

Here's where measurement typically goes wrong:

Mistake 1: Vanity Metrics

Tracking metrics that look impressive but don't correlate with results: "We made 10,000 dials this month!" Great, but did it produce revenue?

Fix: For every metric, ask "does this predict revenue or just make us feel busy?" If it doesn't predict outcomes, stop tracking it.

Mistake 2: Lagging-Only Measurement

Only looking at closed deals and revenue. By which point it's too late to affect outcomes.

Fix: Build leading indicator dashboard that gives early warning of problems while there's still time to fix them.

Mistake 3: Measurement Without Action

Generating reports that no one acts on. Data for data's sake.

Fix: Every metric should have a clear "so what?". If this metric is red, what specific action do we take?

Mistake 4: Analysis Paralysis

Spending more time analysing data than executing based on insights.

Fix: Set time limits on analysis. Make decisions with 80% confidence rather than waiting for 100%. Test, learn, adjust.

Mistake 5: Measuring Individuals in Isolation

Only looking at individual rep performance without considering systemic factors. Territory quality, lead flow, product-market fit in their segment.

Fix: Measure both individual effectiveness and systemic factors. Sometimes low performance is bad execution; sometimes it's bad territory assignment or product issues.

The Measurement Advantage

When you measure strategically with AI enhancement, several things change:

Your forecasts become reliable because you're predicting based on deal health and behavioural patterns, not rep optimism.

Your interventions become timely because leading indicators give early warning when deals need attention.

Your coaching becomes specific because data shows exactly where each rep needs development.

Your process improves continuously because you systematically identify what works and do more of it.

Your team becomes self-correcting because everyone can see their metrics and knows where to improve.

Most importantly, you shift from managing by gut feel to managing by intelligence. Making decisions based on data about what predicts success.

Putting Measurement into Practice

Here's a practical 90-day implementation plan:

Days 1-30: Foundation

- Define 8-12 key metrics (leading and lagging)
- Establish baselines for current performance
- Build basic dashboards (don't let perfect be enemy of good)
- Train team on what metrics mean and why they matter

Days 31-60: Adoption

- Integrate metrics into weekly team meetings
- Begin using leading indicators to drive deal prioritisation
- Start tracking which interventions improve metrics
- Collect feedback on which metrics are useful versus noise

Days 61-90: Optimisation

- Refine metrics based on learnings (remove what's not useful, add what's missing)
- Implement AI-powered predictive analytics if not already in place
- Establish closed-loop process: measure → insight → action → measure improvement
- Celebrate wins driven by data-driven decisions

Notice that the goal isn't perfect measurement from day one. It's building a measurement system that provides useful intelligence and gets progressively better over time.

The Continuous Improvement Loop

The ultimate power of measurement is creating a system that continuously learns and improves:

1. Measure what's happening across your sales process. Activities, behaviours, outcomes.

2. Analyse patterns to understand what predicts success and what predicts failure.

3. Generate insights about where to focus, what to change, what to amplify.

4. Act based on those insights. Coaching, process changes, strategic adjustments.

5. Measure the impact of those actions. Did things improve?

6. Refine and repeat. Keep what worked, discard what didn't, try new interventions.

This loop is what transforms a sales organisation from static (we do things the way we've always done them) to dynamic (we systematically get better every quarter).

AI accelerates every step: it measures more comprehensively, analyses more deeply, generates insights faster, and tracks impact more precisely than humans alone could ever do.

In the next chapter, we'll explore how to build your AI toolkit. Which tools to choose, how to evaluate them, and how to implement them without disrupting your current operations. Because measurement is only possible with the right technology foundation in place. This brings us full circle. The Measure stage feeds back into Qualify, continuously improving how you identify and pursue the right prospects.

CHAPTER OUTCOME

☐ A defined set of leading indicators (the few that truly predict outcomes)

☐ A set of lagging indicators (for confirmation, not steering)

☐ A baseline for your core metrics (current state)

☐ A simple dashboard spec: what you track, how often, and for whom (rep/manager/leader)

☐ A weekly review cadence (agenda + decisions you will make from the data)

☐ A method for turning insights into coaching actions (not just reporting)

☐ A "stop doing" list of vanity metrics that create noise

CHAPTER 10

Building Your AI Toolkit

Selecting and implementing the right tools without drowning in options

---

By the end of this chapter, you will be able to:

- Avoid the tool selection trap by evaluating tech against clear commercial outcomes.
- Apply a tool evaluation framework (problem-solution fit, integration, adoption, data quality, vendor viability, value).
- Identify the essential tool categories for AI-powered sales and what each is responsible for.
- Implement tools in phases (foundation → expansion → optimisation) without breaking your workflows.
- Execute implementation best practices (pilot, training, champions, adoption metrics, iteration).

Two years ago, I watched a promising sales team at a Series B startup nearly collapse under the weight of their own technology stack. They'd embraced AI enthusiastically. Maybe too enthusiastically. In 18 months, they'd adopted 12 different AI-powered sales tools: conversation intelligence, email optimisation, proposal automation, prospecting intelligence, sales engagement, predictive analytics, chatbots, calendar optimisation, and several others I'm probably forgetting.

Each tool had been carefully evaluated. Each promised significant ROI. Each did something valuable. The problem wasn't that any individual tool was bad. It was that collectively, they created chaos.

Sales reps spent their first hour each day just logging into different systems. Data was scattered across platforms with no integration. Insights from conversation intelligence didn't flow to the CRM. Prospecting data didn't sync with email tools. The team was spending 40% of their time on tool administration instead of selling. Turnover spiked as reps grew frustrated with the complexity.

The VP of Sales eventually had to rip out half the stack and start over, this time with a coherent strategy rather than opportunistic tool adoption. Within two quarters, productivity improved dramatically. Not because they had more tools, but because they had the right tools that worked together.

This chapter is about avoiding that mistake. You understand the QUANTUM Framework. You know what capabilities you need. Now you need to make smart decisions about which tools to implement, in what order, and how to integrate them into a coherent ecosystem rather than a chaotic collection of point solutions.

The Tool Selection Trap

At its simplest, the AI sales tool market has exploded. There are now hundreds of vendors claiming to use AI to improve sales outcomes. This abundance creates a paradox: more choice makes good decisions harder, not easier.

Sales leaders face several challenges:

Feature overlap: Many tools claim to do similar things. Five different vendors all promise "AI-powered prospecting." How do you evaluate which is better?

Integration complexity: Tools that don't integrate with each other create data silos and manual work. But evaluating integration quality before purchase is difficult.

Change management: Each new tool requires training, adoption, and behaviour change. Adding too many tools simultaneously overwhelms teams.

Cost accumulation: Individual tools seem reasonably priced ($50-150 per user per month), but 8-10 tools quickly become $800-1,200 per rep per month. A significant investment that needs to show clear ROI.

Vendor stability: The AI sales space is rapidly evolving. Some vendors will be acquired, pivot, or fail. Betting on the wrong vendor creates disruption when you must switch.

The solution isn't avoiding AI tools. They're essential for competitive advantage. It's being strategic about which tools to adopt, when, and how.

The Tool Evaluation Framework

Before evaluating specific tools, establish clear evaluation criteria. Here's the framework I use:

Criteria	Weight	Tool A	Tool B	Tool C	Tool D
tegration	20%	9/10	7/10	8/10	6/10
ity & Accuracy	20%	8/10	9/10	7/10	8/10
rience	15%	7/10	8/10	9/10	7/10
alue	15%	6/10	7/10	8/10	9/10
liability	10%	9/10	8/10	7/10	8/10
y	10%	8/10	9/10	7/10	7/10
ication	10%	7/10	8/10	9/10	6/10
TOTAL		7.75	7.95	7.90	7.25

Guide

= Excellent   7-8 = Good   5-6 = Adequate   1-4 = Needs Work

Figure 10.1: Weighted AI Tool Evaluation Matrix with Scoring Framework - Systematic Assessment Framework for Technology Selection

Criterion 1: Problem-Solution Fit

Question: Does this tool solve a real problem you have, or a theoretical problem vendors claim you should care about?

Start by identifying your biggest constraints:

- Is your pipeline generation insufficient? (Need prospecting tools)
- Are deals stalling in the middle of the cycle? (Need deal intelligence and navigation tools)
- Is forecast accuracy poor? (Need predictive analytics)
- Are reps spending too much time on admin work? (Need automation and efficiency tools)
- Is personalisation at scale impossible? (Need content generation and engagement tools)

Only evaluate tools that address your actual constraints. A brilliant conversation intelligence platform doesn't help if your problem is that reps can't find enough prospects to talk to in the first place.

Criterion 2: Integration and Data Flow

Question: How well does this tool integrate with your existing systems, particularly your CRM?

Evaluate:

- **Native integrations:** Does it have pre-built connections to Salesforce, HubSpot, or whatever CRM you use?
- **Data sync quality:** Does data flow bidirectionally, or only one way? How frequently does it sync?
- **Data quality:** Does integration create duplicate records or data inconsistencies?
- **API robustness:** If you need custom integration, is there a well-documented API?

Poor integration means the tool becomes an island. Valuable in isolation but not contributing to your unified sales ecosystem. This is one of the most common failure points.

Criterion 3: Ease of Use and Adoption

Question: Will your sales reps use this, or will it gather dust after the initial rollout?

Evaluate:

- **Learning curve:** How long does it take a new user to become proficient? Hours, days, or weeks?
- **Daily workflow fit:** Does it integrate into how reps already work, or require completely new behaviours?
- **Mobile accessibility:** Can reps use it effectively from their phones, or is it desktop-only?
- **User experience quality:** Is the interface intuitive, or cluttered and confusing?

The most sophisticated tool is worthless if your team won't use it. Adoption is the difference between ROI and wasted investment.

Criterion 4: Data Quality and Accuracy

Question: Are the data and intelligence this tool provides actually accurate and current?

For prospecting tools: How often is contact data refreshed? What's the accuracy rate for email addresses and phone numbers?

For predictive tools: What's their track record on forecast accuracy? Can they show validation of their models?

For conversation intelligence: How accurate is transcription? Does it correctly identify speakers and key moments?

Bad data is worse than no data. It leads to wrong decisions and wasted effort. Validate accuracy before committing.

Criterion 5: Vendor Viability and Support

Question: Will this vendor still be around and supporting the product in two years?

Evaluate:

- **Company maturity:** Well-funded Series B+ companies are more stable than pre-seed startups
- **Customer base:** Do they have notable customers in your industry and size range?
- **Product roadmap:** Are they actively innovating, or is the product stagnant?
- **Support quality:** What's their support structure? Do they have a customer success team?

You're not just buying software. You're entering a partnership. Make sure the vendor is viable long-term.

Criterion 6: Cost vs. Value

Question: Does the ROI justify the investment, including implementation and ongoing costs?

Calculate total cost:

- Licence fees (typically per user per month)
- Implementation costs (setup, configuration, training)
- Integration costs (IT time or consulting fees)
- Ongoing administration (who manages it day-to-day?)

Compare to expected value:

- Time saved per rep per week
- Increase in conversion rates or deal velocity
- Reduction in cycle time
- Improvement in win rates

If you can't articulate a clear ROI case, don't buy the tool yet.

Essential Tools for AI-Powered Sales

Based on the QUANTUM Framework, here are the tool categories that matter most:

Category 1: Customer Relationship Management (CRM)

Purpose: Central system of record for all customer data, interactions, and opportunities.

Why it matters: Your CRM is the foundation. Everything else integrates with it. If your CRM is weak, your entire tech stack suffers.

Leading options:

- **Salesforce:** Industry standard, most robust, most integrations, most expensive
- **HubSpot:** Strong for SMB and mid-market, great user experience, good value
- **Microsoft Dynamics:** Best for Microsoft-centric environments
- **Pipedrive:** Simpler, more affordable, good for smaller teams

AI capabilities to prioritise:

- Predictive lead scoring
- Opportunity insights and risk flagging
- Automated data entry and enrichment
- Natural language search and reporting

Implementation priority: If you don't have a modern CRM with AI capabilities, this is your first investment. Everything else builds on this foundation.

Category 2: Sales Engagement Platform

Purpose: Automate and optimise outbound prospecting sequences across email, phone, and social.

Why it matters: Enables personalisation at scale for the Tailor stage. Ensures consistent follow-up and timing without manual work.

Leading options:

- **Outreach:** Enterprise-grade, sophisticated workflows, strong analytics
- **SalesLoft:** Similar capability to Outreach, strong customer support
- **Apollo.io:** More affordable, includes prospecting data
- **Groove:** Strong for Salesforce users, simpler interface

AI capabilities to prioritise:

- Email optimisation (subject lines, send time, content suggestions)
- Automated personalisation at scale
- Engagement prediction (who's most likely to respond)
- Cadence optimisation (when to reach out, which channel)

Implementation priority: High for teams doing significant outbound prospecting. Medium if mostly inbound.

Category 3: Conversation Intelligence

Purpose: Record, transcribe, and analyse sales calls to extract insights and coach reps.

Why it matters: Essential for the Understand and Anticipate stages. Captures what happens in conversations so you can learn from it.

Leading options:

- **Gong:** Market leader, most sophisticated analytics, premium pricing
- **Chorus.ai (ZoomInfo):** Strong alternative, good integration with ZoomInfo data

- **Clari Copilot:** Part of broader Clari platform, good for forecasting integration
- **Fireflies.ai:** More affordable option, good basic functionality

AI capabilities to prioritise:

- Accurate transcription and speaker identification
- Key moment detection (objections, commitments, next steps)
- Talk-listen ratio and engagement metrics
- Deal risk prediction based on conversation patterns

Implementation priority: Very high. This is one of the highest-ROI investments you can make.

Category 4: Prospecting and Intelligence

Purpose: Identify and enrich target accounts and contacts with firmographic, technographic, and intent data.

Why it matters: Critical for the Qualify stage. You can't target effectively without good prospect intelligence.

Leading options:

- **ZoomInfo:** Most complete database, strong technographic data
- **Apollo.io:** Good alternative, more affordable, includes engagement tools
- **Clearbit:** Best for real-time enrichment and website visitor identification
- **LinkedIn Sales Navigator:** Essential for relationship intelligence and warm introductions

AI capabilities to prioritise:

- Intent data showing who's actively researching your category
- Technographic intelligence about what tools prospects use
- Relationship mapping showing connections
- Predictive scoring of which accounts to target

Implementation priority: High if doing outbound. Essential for Account-Based strategies.

Category 5: Proposal and Content Management

Purpose: Create, customise, and track proposals and sales content with analytics.

Why it matters: Supports the Tailor stage. Enables personalisation without recreating content from scratch.

Leading options:

- **PandaDoc:** Strong proposal automation, e-signature, analytics
- **Proposify:** Good alternative, more affordable, solid templates
- **Highspot:** Best-in-class for sales content management beyond proposals
- **Seismic:** Enterprise-grade content management and enablement

AI capabilities to prioritise:

- Dynamic content assembly based on prospect profile
- Content recommendations based on deal stage and characteristics
- Engagement analytics (what content prospects view)
- Version control and approval workflows

Implementation priority: Medium-high. More important for complex sales with custom proposals.

Category 6: Forecasting and Pipeline Intelligence

Purpose: Predict deal outcomes and revenue with AI-powered analytics.

Why it matters: Essential for the Measure stage. Enables accurate forecasting and early intervention on at-risk deals.

Leading options:

- **Clari:** Market leader in revenue operations and forecasting
- **Aviso:** Strong AI-powered forecasting and deal insights
- **People.ai:** Good relationship intelligence and pipeline analytics
- **InsightSquared:** More affordable option with solid analytics

AI capabilities to prioritise:

- Deal-level risk scoring and probability prediction
- Pattern-based forecasting (not just rep input)
- Anomaly detection for deals that need attention
- What-if scenario modelling for pipeline management

Implementation priority: High for larger teams ($5M+ ARR). Medium for smaller organisations.

The Phased Implementation Approach

Don't try to implement everything at once. Here's a sensible phasing:

Phase 1: Foundation (Months 1-3)

Focus: Get core systems right before adding complexity.

Priority tools:

- **CRM (if not already in place or needs upgrade):** Everything builds on this
- **Conversation Intelligence:** Highest ROI, immediate coaching value
- **Sales Engagement OR Prospecting Intelligence:** Choose based on whether your constraint is execution (engagement) or targeting (prospecting)

Goal: Establish data foundation and capture all sales activities and conversations in one place.

Phase 2: Expansion (Months 4-6)

Focus: Add capabilities that apply your foundation.

Priority tools:

- Whichever you didn't add in Phase 1 (Engagement or Prospecting)
- **Proposal/Content Management** (if complex sales)
- **Email optimisation tools** (if doing high-volume outbound)

Goal: Enable personalisation at scale and improve activity effectiveness.

Phase 3: Optimisation (Months 7-12)

Focus: Add sophisticated analytics and advanced capabilities.

Priority tools:

- **Forecasting and Pipeline Intelligence**
- **Advanced integrations** between existing tools
- **Specialized tools** for specific use cases (ABM platforms, pricing optimisation, etc.)

Goal: Mature analytics that enable continuous improvement and predictive capabilities.

The Integration Architecture

As you build your stack, think architecturally about how tools connect:

Hub-and-Spoke Model

Your CRM is the hub. All other tools are spokes that connect to it. This keeps CRM as the system of record while specialized tools add capabilities.

Advantages: Simple model, clear data ownership, most tools have CRM integrations

Disadvantages: CRM becomes bottleneck, not all workflows fit CRM-centric model

Best-of-Breed with iPaaS

Use an integration platform (Zapier, Workato, Tray.io) to connect tools that don't natively integrate well.

Advantages: More flexibility, can connect any tools, sophisticated workflow orchestration

Disadvantages: Another platform to manage, can become complex, creates dependency

Platform Play

Choose tools from vendors building complete platforms (like Salesforce ecosystem or HubSpot ecosystem) rather than best-of-breed point solutions.

Advantages: Guaranteed integration, unified experience, single vendor relationship

Disadvantages: Potentially not best-in-class for specific functions, vendor lock-in

Most successful teams use a hybrid: CRM as hub, native integrations where available, iPaaS for gaps, and platform approach for core workflows.

Implementation Best Practices

Here's how to roll out tools successfully:

Start with a Pilot

Don't roll out to entire team immediately. Start with 5-10 reps for 30-60 days. Learn what works, identify issues, refine before broad deployment.

Invest in Training

Budget 2-3 hours of initial training per tool, plus ongoing coaching. Tools fail due to poor adoption far more often than poor functionality.

Establish Champions

Identify 2-3 power users who embrace new tools. They become peer coaches who help others adopt and troubleshoot.

Measure Adoption and Impact

Track: What percentage of reps actively use the tool? Is the metric you bought it to improve improving? If adoption is low or impact isn't clear after 90 days, investigate why.

Iterate Based on Feedback

Collect user feedback continuously. Some tools will work great. Others will need configuration adjustments or workflow changes. Some might need to be replaced.

Common Tool Selection Mistakes

Here's where teams typically go wrong:

Mistake 1: Tool-First Rather Than Problem-First

Buying tools because they're hot or competitors have them, not because they solve your specific problems.

Fix: Always start with "what's our constraint?" then find tools that address it.

Mistake 2: Ignoring Integration

Choosing best-of-breed tools that don't integrate well, creating data silos.

Fix: Make integration quality a primary evaluation criterion, not an afterthought.

Mistake 3: Underestimating Change Management

Assuming reps will automatically adopt new tools without training and reinforcement.

Fix: Budget time and resources for training, coaching, and ongoing adoption support.

Mistake 4: Over-Complicating Too Quickly

Adding too many tools simultaneously, overwhelming team with complexity.

Fix: Implement in phases. Get each tool working well before adding the next.

Mistake 5: Not Measuring ROI

Continuing to pay for tools that aren't delivering measurable value.

Fix: Establish success metrics before purchase. Review quarterly. Eliminate tools that aren't paying off.

Your 90-Day Action Plan

Here's a practical plan to build your toolkit:

Days 1-30: Assessment

- Audit current tools and usage (what do you have, is it being used effectively?)
- Identify your top 3 constraints (pipeline generation, deal progression, forecast accuracy, etc.)
- Research tools that address those constraints
- Create shortlist (2-3 options per category you need)

Days 31-60: Evaluation

- Schedule demos with shortlisted vendors
- Run proof-of-concept or trial with top choice in each category
- Test integration with your existing systems
- Gather feedback from pilot users
- Make purchase decisions

Days 61-90: Implementation

- Deploy first tool to pilot group
- Train users and establish best practices
- Monitor adoption and gather feedback
- Refine configuration and workflows
- Plan rollout to full team

Then repeat the cycle for next tool category, learning from first implementation.

The Toolkit Advantage

When you've built the right AI toolkit with the QUANTUM Framework in mind:

Your reps spend more time selling because tools handle research, data entry, and administrative work.

Your decisions improve because you have better intelligence at every stage of the sales process.

Your personalisation scales because AI handles mechanical customisation while humans add strategic judgment.

Your forecasts are accurate because predictive analytics replace gut feel.

Your team continuously improves because conversation intelligence and analytics reveal what works.

Most importantly, you've created a competitive advantage that compounds over time. While competitors are drowning in their technology or still working manually, you're using AI systematically to outperform.

In the next chapter, we'll explore advanced strategies for using this toolkit. Account-Based Selling approaches, complex enterprise deal orchestration, and sophisticated techniques that separate elite performers from the rest of the pack.

CHAPTER OUTCOME

☐ A clear list of the outcomes you want tools to deliver (not features you want to buy)

☐ A tool evaluation scorecard (fit, integration, adoption, security, data, ROI, vendor viability)

☐ A short list of must-have tool categories for your context (CRM, enablement, intelligence, automation, analytics, etc.)

☐ A phased implementation plan (what comes first, second, third. And why)

☐ A pilot plan (who, duration, success criteria, measurement)

☐ An adoption plan (training, champions, incentives, enablement)

☐ A "don't break the workflow" principle: what cannot become more complex

CHAPTER 11

Advanced Strategies

Elite techniques for complex deals and strategic accounts

By the end of this chapter, you will be able to:

- Decide when account-based selling is the right motion and how to run strategic account orchestration.
- Use competitive intelligence to shape positioning and apply "kill strategies" ethically and effectively.
- Work through complex enterprise mechanics (matrix orgs, gatekeepers, exec committees, alliances).
- Shift from vendor to partner through co-creation, joint value, and multi-year planning.
- Apply advanced personalisation tactics (predictive, ecosystem-based, timeline-based) in high-stakes deals.

Five years into my sales career, I thought I understood complex selling. I'd closed seven-figure deals, navigated enterprise decision-making, and consistently hit quota. Then I joined a company where the average deal cycle was 18 months, involved 15+ stakeholders, and required coordination across multiple business units and geographies. My previous "complex" deals suddenly seemed simple by comparison.

I quickly learned that sophisticated sales scenarios require strategies that go beyond the fundamentals. The QUANTUM Framework provides the foundation. But elite execution in high-stakes environments requires additional capabilities: orchestrating account-based strategies across multiple opportunities, threading influence through matrixed organisations, managing competitive dynamics in zero-sum evaluations, and coordinating internal resources across extended timelines.

In effect, this chapter is about those advanced strategies. Not every seller needs them. If you're closing transactional deals in 30-60 days with 2-3 stakeholders, the core QUANTUM Framework is sufficient. But if you're pursuing strategic accounts, enterprise deals, or transformational sales, these techniques separate those who consistently win from those who occasionally get lucky.

Account-Based Selling: The Strategic Account Approach

Traditional sales are opportunity-centric: you qualify opportunities one at a time, work them individually, and either win or lose each deal independently.

Account-Based Selling (ABS) flips this model: you identify strategic target accounts, pursue them end-to-end across multiple potential opportunities, and build relationships that transcend any single transaction.

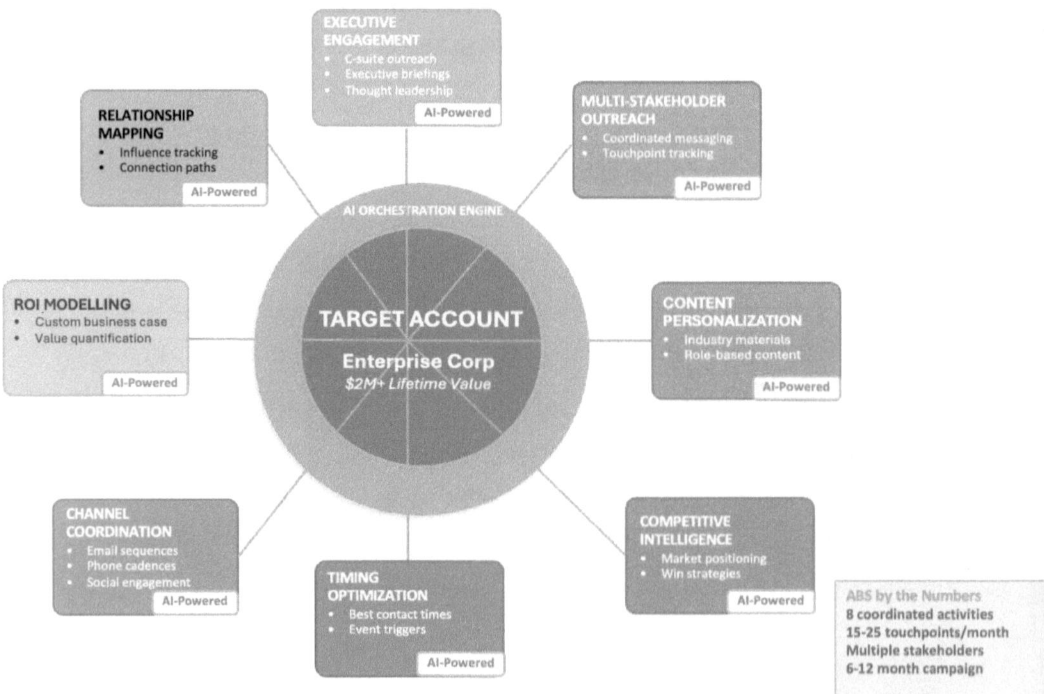

Figure 11.1: AI-Orchestrated Account-Based Selling - Simultaneous Multi-Vector Engagement
AI-Coordinated Multi-Touch Strategic Account Engagement

When Account-Based Selling Makes Sense

ABS is not for everyone. It requires significant investment per account, which only makes sense when:

Deal sizes justify the effort: If your typical deal is $10K, you can't afford to spend months cultivating an account. But if you're pursuing $500K+ annual contracts with $2M+ lifetime value, intense focus on the right accounts pays off.

Account expansion potential is high: The ideal ABS target isn't just one large initial deal. It's an account where you can expand from one department to enterprise-wide, one-use case to many, or initial purchase to strategic partnership.

Long sales cycles allow relationship building: If prospects buy in 30 days, you don't have time for strategic account development. But in 6-12+ month cycles, relationship investments compound.

Decision-making is complex and political: When deals involve 10+ stakeholders, executive committees, and organisational politics, generic outreach won't work. You need strategic, coordinated engagement.

Market concentration favours focus: If your ideal customer base is 500 large enterprises rather than 50,000 small businesses, targeting specific accounts makes more sense than broad-based prospecting.

The ABS Strategic Framework

Account-Based Selling through the QUANTUM lens:

Qualify (Account Selection): Rather than qualifying individual opportunities, you qualify accounts for strategic pursuit. Criteria include:

- **Strategic fit:** Perfect match to your ICP with high lifetime value potential

- **Market importance:** Reference value or strategic partnership potential
- **Penetration opportunity:** Currently underserved by competitors
- **Executive access:** Potential for C-level relationships
- **Expansion potential:** Multiple business units or use cases to pursue

Select 10-25 accounts that meet these criteria. This focus is intentional. You're investing significant resources per account.

Understand (Deep Account Intelligence): Go far beyond typical research:

- Map the complete organisational structure, not just your target department
- Understand their strategic initiatives, not just tactical pain points
- Identify all potential entry points and expansion opportunities
- Research key executives' backgrounds, priorities, and decision-making styles
- Analyse their technology ecosystem and vendor relationships
- Monitor their market activity, financial performance, and strategic direction

This isn't one-time research. It's ongoing intelligence gathering that builds over months.

Anticipate (Strategic Scenario Planning): Predict not just objections, but:

- Which executives will champion technology investments vs. resist them
- When budget cycles create buying windows
- How organisational changes (new leaders, restructuring) create opportunities
- Which competitive dynamics might emerge

- How macroeconomic or industry trends affect their priorities

Steer through (Multi-Threaded Orchestration): In ABS, navigation becomes orchestration:

- Build relationships across multiple departments simultaneously
- Coordinate engagement between your executives and theirs
- Orchestrate "surround sound". The account hears about you from multiple sources
- Create peer connections between their teams and your customer references

You're not just selling one deal. You're becoming embedded in the account.

Tailor (Hyper-Personalisation): Every interaction is customised:

- Executive briefings tailored to specific C-level priorities
- Department-specific demonstrations addressing unique workflows
- Custom thought leadership addressing their industry challenges
- Personalised content for each key stakeholder

Unify (Coordinated Account Team): ABS requires internal coordination:

- Account Executive owns the relationship and strategy
- Solutions Engineer provides technical expertise
- Customer Success previews post-sale value
- Executive sponsor engages with their executives
- Marketing creates custom content and events

Everyone coordinates around a unified account plan.

Measure (Account Health Metrics): Track account-level indicators:

- Relationship depth and breadth (how many relationships, at what levels)
- Engagement velocity (is interest growing or declining)

- Competitive positioning (where do we stand vs. alternatives)
- Pipeline development (are opportunities emerging)
- Executive sponsorship strength (do we have C-level champions)

The Account-Based Orchestration Model

Here's how coordinated ABS works in practice:

Research phase (Months 1-2): Deep intelligence gathering using AI tools to understand the account comprehensively. Build detailed account maps showing stakeholders, priorities, technology environment, and potential opportunities.

Engagement phase (Months 3-4): Multi-channel outreach:

- Executive to executive: Your CEO/VP reaches out to their C-level
- Subject matter experts: Your product leaders connect with their technical teams
- Thought leadership: Custom content addressing their specific industry challenges
- Events: Invite them to exclusive roundtables or executive briefings
- Warm introductions: Use mutual connections and customers

Discovery phase (Months 5-7): Deep consultative engagement:

- Understand strategic initiatives across multiple departments
- Identify pain points and opportunities across the organisation
- Co-create vision for how you could partner strategically
- Build business cases for multiple potential use cases

Opportunity development (Months 8-12): Convert relationships into opportunities:

- Start with one or two highest priority use cases
- Use initial success to expand into additional departments
- Build champion network that advocates internally

- Manage procurement and contracting as strategic partner, not transactional vendor

Expansion (Ongoing): After initial win, systematically expand:

- Additional use cases within initial department
- Same use case in other departments
- Cross-sell adjacent products
- Evolve from vendor to strategic partner

Notably, the timeline varies, but the pattern is consistent: strategic, patient, coordinated pursuit that builds momentum over time.

Competitive Strategy in Complex Sales

In sophisticated enterprise sales, you're rarely the only option. Managing competitive dynamics requires strategies beyond "we're better than them."

Competitive Intelligence

Use AI tools to monitor competitive activity:

Conversation intelligence flags competitor mentions in sales calls automatically: "Competitor X was mentioned 3 times in today's call with Acme Corp. Analysis suggests they're ahead in technical evaluation but prospect has concerns about support."

Intent data platforms show when prospects research competitors: "Acme Corp has increased research activity on Competitor Y's website by 300% in past 2 weeks."

Social listening tracks when prospects engage with competitor content: "VP of Sales at Acme Corp downloaded Competitor Z whitepaper and attended their webinar."

This intelligence lets you address competitive threats proactively rather than being surprised.

Competitive Positioning Strategies

How you position against competition depends on the situation:

Head-to-head (when you're clearly better): Emphasise direct comparison. "Here's how we stack up on the 5 most important criteria." Use comparison matrices, third-party evaluations (G2, Gartner), and customer references who switched from competitors.

Differentiation (when you're different, not necessarily better): Position yourself in a different category. "They're the enterprise platform for companies that want complete features. We're the specialized solution for companies that prioritise [your strength]." Make it apples-to-oranges comparison where you win on the criteria that matter to this buyer.

Below-the-line (when incumbent is entrenched): Don't attack the incumbent directly. They have relationships and switching costs on their side. Instead, position for a specific use case or department where you can win, then expand. "We understand you're using Incumbent X enterprise-wide. We're not suggesting you rip that out. But for this specific new initiative, here's why starting with us makes sense."

FUD avoidance (when competitor is spreading fear, uncertainty, doubt): Address competitive attacks directly but professionally. "You may hear competitors claim we can't handle enterprise scale. Here are 15 Fortune 500 customers who've proven otherwise." Don't be defensive but do counter misinformation factually.

The Competitive Kill Strategy

When you need to actively displace a competitor:

Step 1: Identify their weakness. Every vendor has gaps. Use conversation intelligence to discover what prospects complain about regarding competitors: poor support, difficult implementation, limited flexibility, high cost.

Step 2: Position that weakness as deal critical. Make the criteria where they're weak seem essential: "Implementation time is critical for you given your timeline. Here's data showing Competitor X's implementations typically take 2x longer than ours."

Step 3: Provide proof. Customer references who switched from that competitor specifically. "Tech Corp switched from Competitor X to us specifically because of implementation speed. Happy to connect you with their VP of Sales."

Step 4: Create urgency around their weakness. "If you move forward with Competitor X, you're looking at Q3 go-live at earliest. With us, you could be live by end of Q1. Given your business objectives, that 6-month difference is significant."

Notice how this isn't about being dishonest. It's about making real competitive differences salient to the decision.

Complex Enterprise Navigation

Enterprise sales often involve managing organisational complexity that makes mid-market sales look simple:

Matrix Organisations

In matrixed environments, authority is shared across functional departments and business units. The VP of Sales might want your solution, but needs buy-in from IT, Finance, Procurement, Legal, and regional business unit leaders.

Strategy:

- Map the decision-making matrix explicitly: who has input, who has approval, who has veto
- Build relationships in parallel across multiple dimensions (functional and business unit)
- Create cross-functional alignment by showing value to each stakeholder group

- Identify and enable a "chief coordinator" internally who navigates the matrix for you

AI support: Use relationship intelligence tools to map the organisational network. "These 5 people all report to different executives but work together regularly. Getting alignment among them is critical."

Procurement and Legal Gatekeepers

Large organisations have procurement and legal teams that can kill deals late in the cycle with contract requirements you can't meet.

Strategy:

- Engage procurement early, not just when it's time to sign contracts
- Understand their priorities (cost savings, vendor management, risk mitigation)
- Position yourself as making their job easier, not harder
- Pre-negotiate terms on key issues before you're in final negotiations with limited use
- Build business case that includes procurement's success metrics

Common pitfalls: Treating procurement as obstacle rather than stakeholder. Assuming business stakeholders' enthusiasm overrides procurement concerns. Waiting until contract stage to address legal issues.

Executive Committee Approval

Some purchases require executive committee or board approval. This creates challenges because you can't attend those meetings. Your champion presents on your behalf.

Strategy:

- Arm your champion with executive-ready materials: one-page summary, business case, risk mitigation plan

- Anticipate questions executives will ask strategic alignment, financial justification, competitive implications, implementation risk
- Role-play the executive presentation with your champion
- Provide comparison to alternatives (including status quo and internal build)
- Make it easy for champions to advocate effectively without your presence

Mistake to avoid: Assuming your champion knows how to sell internally. They're often less experienced at internal selling than you are at external selling. Coach them explicitly.

Strategic Partnerships and Alliances

The most sophisticated sales aren't vendor-customer relationships. They're strategic partnerships that create mutual value.

From Vendor to Partner

Moving beyond transactional sales:

Phase 1: Vendor (Transactional): You provide product, they pay. Relationship is arms-length. Easily replaceable.

Phase 2: Trusted Vendor (Preferred): You've proven value. They prefer you but still evaluate alternatives. Relationship is professional but limited.

Phase 3: Strategic Supplier (Integrated): You're embedded in their operations. Switching costs are high. Relationship includes multiple stakeholders and touch points.

Phase 4: Strategic Partner (Collaborative): You co-create value. They see you as extension of their team. Joint roadmaps, integrated planning, shared success metrics. Extremely difficult to displace.

Getting to Phase 4 requires:

- Delivering exceptional value consistently
- Aligning your roadmap with their strategic priorities
- Executive relationships that go beyond sales to genuine partnership
- Demonstrating you care about their success, not just your revenue
- Investing in their success proactively (training, best practices, dedicated resources)

Co-Creation and Joint Value

True partnerships involve mutual value creation:

Joint innovation: Work together to develop new capabilities that benefit both parties. "We built this feature specifically based on your use case. You get early access and influence the roadmap. We get a marquee customer and real-world validation."

Thought leadership: Co-author content, present together at conferences, create case studies that showcase both organisations.

Reference and advocacy: They become public advocates, not just references. They speak at your events, quote in your marketing, help you win other customers.

Commercial creativity: Move beyond transactional pricing to alignment models where you both win when they succeed. Revenue share, performance-based pricing, or joint venture structures.

Multi-Year Strategic Account Planning

The most advanced ABS involves long-term account planning:

Year 1: Establish beachhead. Win initial opportunity, prove value, build foundational relationships.

Year 2: Expand across organisation. Use initial success to penetrate additional departments or use cases. Deepen executive relationships.

Year 3: Become strategic partner. Align roadmaps, create custom solutions, establish executive-to-executive relationships that transcend individual deals.

Year 4+: Evolve to enterprise-wide standard. Displace remaining competitors, become deeply embedded, shift to renewal and expansion focus rather than new sales.

This requires patience and organisational commitment. You're optimising for lifetime value, not quarterly revenue.

Advanced Personalisation Techniques

Beyond the Tailor strategies in Chapter 7, elite sellers use sophisticated personalisation:

Predictive Personalisation

Use AI to predict not just what prospects care about now, but what they'll care about next:

"Based on patterns from similar companies, 3-6 months after implementing your current initiative, you'll likely face [new challenge]. Here's how our roadmap aligns with addressing that when you're ready."

This demonstrates you understand their journey, not just their current state.

Ecosystem Personalisation

Tailor not just to the individual or company, but to their entire business ecosystem:

- Reference customers and partners they work with
- Show how you integrate with their specific tech stack
- Address their industry's unique regulatory or competitive dynamics

- Demonstrate understanding of their supply chain or distribution channels

Consider that this shows you understand their world, not just their business.

Timeline Personalisation

Adapt your engagement to their buying timeline without pushing:

Early stage (not actively buying): Educational content, thought leadership, relationship building. No pressure.

Middle stage (actively evaluating): Detailed proof points, technical validation, business case development. Helpful but not pushy.

Late stage (decision imminent): Urgency, clear differentiation, risk mitigation, final objection handling. Assertive but respectful.

AI can detect timeline stage based on engagement patterns and adjust messaging automatically.

Crisis and Opportunity Management

Advanced sellers know how to respond to unexpected events:

Competitive Displacement Urgency

When a competitor wins a deal at a strategic account, it's a crisis. But also an opportunity:

Immediate: Get post-mortem intelligence. Why did they choose the competitor? What could you have done differently?

Short-term (0-6 months): Stay engaged. Implementation often goes poorly. Be ready when buyer's remorse sets in.

Medium-term (6-18 months): Position for expansion opportunities the competitor didn't win. "We understand they won the initial deal. When you're ready to expand to [adjacent use case], we'd love to discuss."

Long-term (18+ months): Contract renewal. Build relationships for when their contract comes up for renewal.

Leadership Change Opportunity

New executives create buying windows:

New CEO: Brings strategic priorities that create budget for initiatives

New CTO: Often wants to make their mark with technology modernisation

New CFO: May prioritise efficiency and ROI differently than predecessor

New business unit leader: Wants to prove themselves with quick wins

Monitor for these changes using AI tools and reach out quickly with relevant value propositions aligned to what new leaders typically prioritise.

Market Event Response

Industry disruptions create opportunities:

Regulatory changes: Position your solution as compliance enabler

Competitive consolidation: Position yourself as alternative to newly merged giant

Market downturns: Emphasise ROI and cost savings

Technology disruptions: Show you're aligned with future, not past

Elite sellers have playbooks for responding to these events quickly and relevantly.

Putting Advanced Strategies into Practice

You don't need every advanced strategy immediately. Here's how to graduate to sophisticated techniques:

Master the fundamentals first. Don't try Account-Based Selling if you haven't mastered basic QUANTUM execution. Advanced strategies amplify good fundamentals; they don't compensate for weak basics.

Start with one strategic account. Pick your highest-value target account and apply ABS rigorously. Learn what works. Then expand to 5 accounts, then 10-25.

Build internal capabilities gradually. Advanced strategies require coordination, executive engagement, and specialized skills. Develop this over time rather than trying to transform overnight.

Measure and refine. Track whether advanced techniques are improving win rates, deal sizes, and expansion revenue. If not, refine your approach.

Worth noting: the goal isn't complexity for its own sake. It's deploying sophisticated techniques when they're warranted by deal size, strategic importance, and competitive dynamics.

In the next chapter, we'll address the ethical dimensions of AI-powered sales. How to maintain authenticity, respect customer privacy, and build trust while putting to work powerful technology that could be misused in the wrong hands.

CHAPTER OUTCOME

☐ A decision: which deals require advanced orchestration vs standard motion

☐ A strategic account plan structure (goals, stakeholders, value hypothesis, actions, risks)

☐ A competitive intelligence routine (what to monitor and how to act on it)

☐ A plan for executive engagement (when to bring execs in and what to ask them to do)

☐ A play for complex enterprise dynamics (committees, procurement, risk, compliance, legal)

☐ A "partner posture" plan: how you shift from vendor to co-creator

☐ An escalation plan for when deals stall (what changes immediately)

CHAPTER 12

Ethics and Best Practices

By the end of this chapter, you will be able to:

- Distinguish ethical persuasion from manipulation and apply a five-question test to any AI-driven sales tactic.
- Set clear privacy boundaries for prospect research—knowing what data to use, what to avoid, and why.
- Define an authenticity standard for AI-generated content (when to disclose, when to add personal voice, when to override).
- Build transparency and consent practices around call recording, data usage, and AI-assisted communication.
- Design an ethical sales culture with leadership standards, balanced incentives, and technology governance.

Using AI responsibly while maintaining trust and authenticity

About three years ago, I received an email from a vendor that made me deeply uncomfortable. It was personalised. Impressively so. It referenced specific projects I was working on, challenges my team faced, even a recent LinkedIn post I'd made. The research was thorough. The relevance was undeniable.

But something felt wrong. The personalisation crossed a line from "I've done my homework" to "I've been watching you." Details that could only have come from monitoring my digital footprint in ways that felt invasive. It was effective research weaponized in a way that felt manipulative rather than helpful.

I never responded. More importantly, I told colleagues about the experience. And not in a positive way. That vendor burned their reputation with me and several others through aggressive use of technology that prioritised tactics over trust.

This experience crystallised something I'd been thinking about as AI tools became more powerful: just because you can do something doesn't mean you should. The technology enables personalisation, prediction, and persuasion at levels that were impossible just a few years ago. But with that capability comes responsibility.

This chapter addresses the ethical dimensions of AI-powered sales. These aren't abstract philosophical questions. They're practical issues that affect your reputation, your relationships, and ultimately your long-term success. The salespeople who thrive in the AI era won't just be those who use the technology most aggressively. They'll be those who use it most responsibly.

The Ethics Challenge in AI-Powered Sales

AI creates several ethical tensions that didn't exist in traditional sales:

Information asymmetry: You can know far more about prospects than they realise. Where's the line between research and surveillance?

Authenticity vs. automation: When AI generates your messages, are you being authentic or deceptive? Does it matter if the communication is helpful?

Manipulation risk: AI can identify psychological patterns and optimise messaging for persuasion. When does persuasion become manipulation?

Privacy boundaries: How much personal information should you tap into? Just because it's publicly available doesn't mean using it is appropriate.

Transparency obligations: Should you disclose when AI generates content, analyses conversations, or scores prospects? Or is that irrelevant as long as you're providing value?

Consider that these aren't questions with simple answers. But ignoring them leads to practices that might work short-term but damage trust and reputation long-term.

The Foundation: Trust as Long-Term Strategy

Before we dive into specific ethical issues, let's establish why ethics matters beyond just "it's the right thing to do":

Trust is your competitive advantage. In markets where products and prices are increasingly similar, trust differentiates. Prospects buy from people they trust. That trust can't be built through manipulation.

Reputations are fragile. One viral LinkedIn post about creepy sales tactics can damage a reputation it took years to build. Your digital footprint is permanent.

Regulatory trends favour privacy. GDPR in Europe, CCPA in California, and emerging privacy regulations globally are making aggressive data practices illegal, not just unethical.

Customers are getting savvier. People increasingly understand when they're being manipulated by algorithms. Practices that work today may backfire tomorrow as prospects become more sophisticated.

Internal culture matters. Sales teams that operate ethically have lower turnover, higher morale, and better long-term performance. Cultures that prioritise winning at any cost burn people out.

The most successful long-term sales organisations aren't those that exploit every possible advantage. They're those that build trust systematically by doing the right thing even when no one's watching.

Privacy: Where to Draw the Line

Let's start with the most immediate ethical question: how much prospect information should you gather and use?

The Public vs. Private Distinction

Information falls into three categories:

Clearly public: Professional information people explicitly share publicly. LinkedIn profiles, company websites, press releases, conference presentations, blog posts. Using this is unambiguously appropriate.

Technically public but contextually private: Information that's technically accessible but not intended for broad consumption. Personal social media not marked private, photos from public events, old blog posts or comments from years ago. Legal to access, but ethically questionable to use.

Private: Information behind login walls, shared in private groups, or obtained through data breaches or deceptive practices. Using this is both unethical and often illegal.

The ethical principle: Use information people would reasonably expect you to access and apply professionally. If someone would be uncomfortable knowing you found and used certain information, that's a signal you've crossed a line.

Personal vs. Professional Information

Just because information is public doesn't mean it's professionally appropriate to use:

Generally appropriate:

- Professional background and experience
- Published work and thought leadership
- Company role and responsibilities
- Professional interests and expertise
- Public company information and challenges

Ethically questionable:

- Personal hobbies unrelated to business context
- Family information (spouse, children)
- Personal photos or social media
- Political or religious affiliations
- Health information

In practice, the test: Would referencing this information in outreach make the prospect feel:

- **"They understand my professional world"** (appropriate)
- **"They've done impressive research"** (appropriate)
- **"That's creepy. How do they know that?"** (inappropriate)

If you wouldn't want to explain how you found information in a face-to-face conversation, don't use it in digital outreach.

AI and Privacy Boundaries

AI tools can aggregate information from dozens of sources automatically. This creates capability that requires judgment:

Just because AI can compile a thorough dossier doesn't mean you should use all of it. Apply filters:

- Does this information help me serve the prospect better, or just appear knowledgeable?
- Would the prospect feel respected if they knew how thoroughly I'd researched them?

- Am I using information to create value or to create undue pressure?

Example of appropriate use: "I noticed your company recently expanded into European markets based on your press releases. Most companies at your stage face challenges with international sales coordination. Is that consistent with what you're experiencing?"

Example of inappropriate use: "I saw on Facebook you just got back from vacation in Italy. I noticed your company is expanding into Europe..."

The first demonstrates professional research relevant to the conversation. The second demonstrates surveillance that contributes nothing except discomfort.

Authenticity in the Age of AI-Generated Content

When AI generates your emails, proposals, and messages, a fundamental question arises: are you being authentic?

The Authenticity Spectrum

Let's distinguish between different levels of AI assistance:

AI as tool (clearly authentic): You write content, AI checks grammar, suggests better phrasing, or offers alternative word choices. This is like spell-check on steroids. Clearly your content, just polished.

AI as assistant (generally authentic): You provide direction and context, AI generates a draft, you review and refine significantly. The final content reflects your thinking even though AI did initial drafting. This is like having a junior writer who drafts based on your outline.

AI as ghost-writer (potentially problematic): AI generates content with minimal human input or review. You might change a few words but essentially send what AI wrote. This raises authenticity questions.

AI as impersonator (ethically problematic): AI generates content designed to mimic your personal style so closely that recipients can't tell the difference. Even if content is helpful, the deception is concerning.

The Disclosure Question

Should you disclose when AI helps create content? There's no universal answer, but here's a framework:

Don't need to disclose:

- Using AI for research and information gathering
- AI-assisted editing and proofreading
- AI-generated insights that you validate and personalise
- Using AI as brainstorming tool that you refine significantly

Consider disclosing:

- When AI generates substantial content, you minimally edit
- In situations where authenticity is explicitly valued (personal notes, thank-you messages)
- When asked directly about your process

Should disclose:

- When creating content that purports to be personally handcrafted
- In contexts where personal attention is the value proposition
- If not disclosing would constitute deception

The principle: Focus less on whether AI was involved and more on whether the final content authentically represents your thinking and provides real value. A thoughtful AI-assisted message is more authentic than a thoughtless manually written template.

Maintaining Your Voice

In practice, the risk with AI-generated content isn't that it's inauthentic per se. It's that it's generic. To maintain authenticity:

Develop distinct AI prompts that reflect your style: Train AI on your actual writing so generated content sounds like you, not like generic corporate speak.

Always add personal touches: Even if AI drafts 90%, you should add specific references, personal observations, or unique insights that make it unmistakably yours.

Use AI for structure, not soul: Let AI handle organisation, formatting, and polished phrasing. You provide strategic direction, specific insights, and genuine connection.

Validate factual accuracy: AI sometimes hallucinates or gets details wrong. Everything AI generates should be fact-checked before sending.

Manipulation vs. Persuasion

AI can identify psychological patterns, optimise messaging for conversion, and personalise approaches based on personality analysis. This raises questions: when does persuasion become manipulation?

The Ethical Distinction

Persuasion (ethical): Presenting your solution's genuine value in ways that hit home with the prospect's priorities, using language and framing that makes the value clear. You're helping them make a good decision.

Manipulation (unethical): Using psychological techniques to push prospects toward decisions not in their interest, exploiting cognitive biases or emotional vulnerabilities for your benefit rather than theirs.

The key difference: Intent and outcome alignment.

- **Persuasion:** You genuinely believe your solution serves their needs and you're helping them understand that value.
- **Manipulation:** You're prioritising your sale over their wellbeing, using techniques that exploit rather than inform.

Red Flags for Manipulation

You're crossing into manipulation when you:

Exploit urgency artificially: Creating false scarcity or pressure that doesn't reflect reality. "This price is only available today" when it's available all week.

Deliberately obscure important information: Hiding costs, limitations, or requirements to avoid losing the deal, planning to "clarify" after commitment.

Target known vulnerabilities: Using personality analysis to exploit people's insecurities or fears rather than address genuine needs.

Misrepresent alternatives: Deliberately mischaracterising competitors or status quo to make your solution seem better than it is.

Pressure during weakness: Targeting prospects when they're under stress, facing crisis, or making emotional decisions rather than thoughtful ones.

The Transparency Test

Would you be comfortable explaining your techniques if prospects knew about them?

Passes the test: "I researched your company to understand your challenges so I could show you how we've helped similar companies. Here's what I found..."

Fails the test: "I used AI to analyse your personality type so I could craft messages that trigger your specific psychological patterns..."

The first is helpful research. The second is manipulative psychology.

The Truth Imperative

This should be obvious, but AI makes certain forms of deception easier, so it's worth stating explicitly:

Never lie or misrepresent facts. Not about your product, not about competitors, not about customer results, not about anything. The sophistication of AI doesn't change this fundamental principle.

AI-specific temptations to resist:

Fake personalisation: Using AI to create the illusion you've personally researched someone when you haven't. "I loved your recent article on [topic]" when AI generated that line, and you've never read the article.

False authority: Citing AI-generated "statistics" or "studies" that don't exist. AI sometimes makes up convincing-sounding data. Always verify.

Implied relationships: Using AI to suggest connections or references you don't have. "People in your industry like [Name] love our solution" when Name isn't a customer.

Misleading capabilities: Claiming AI features you don't have because prospects expect AI functionality. Be honest about what your AI does and doesn't do.

Manufactured urgency: Using AI to create false scarcity or pressure. "AI predicts inventory will run out" when inventory is fine.

The Correction Obligation

When mistakes happen (and with AI, they will), you have an ethical obligation to correct them promptly:

If AI generates factually incorrect information that you send: Correct it immediately and apologize.

If you discover you've misrepresented something: Clarify truthfully, even if it's uncomfortable.

If you realise you crossed an ethical line: Acknowledge it and change behaviour going forward.

Your reputation is built on how you handle mistakes as much as avoiding them in the first place.

Informed Consent and Transparency

Do prospects have a right to know how you're using AI in your interactions with them?

Recording and Analysis Disclosure

When recording sales calls for conversation intelligence:

Legal requirements: Many jurisdictions require all-party consent for recording. Comply with laws, obviously. "This call is being recorded" isn't just courtesy. It's often legally required.

Ethical best practice: Even when not legally required, disclose that calls are recorded and analysed. "We record our calls for training and quality purposes" is standard and appropriate.

What you don't need to disclose: The specific AI techniques you're using to analyse calls. Prospects don't need to know you're using sentiment analysis or talk-time ratios.

AI-Generated Content Disclosure

As discussed earlier, you generally don't need to disclose that AI assisted with drafting content. But there are exceptions:

When authenticity is the value proposition: If you position yourself as providing white-glove, personally crafted service, using AI extensively without disclosure is arguably deceptive.

When directly asked: If a prospect asks, "Did you write this yourself?" answer truthfully.

When the relationship warrants it: With close, long-term relationships where trust is paramount, being transparent about your process (including AI assistance) can strengthen the relationship.

Data Usage Transparency

Prospects should have reasonable understanding of what you do with their data:

Appropriate: "We use AI to analyse our sales conversations to improve how we serve customers." (True and helps them)

Inappropriate: Not disclosing that you're tracking their website behaviour, social media activity, and email engagement to score their likelihood of purchase. (Legal in many places, but feels invasive if not disclosed)

Best practice: Privacy policies and terms that clearly explain data usage, even if most people don't read them.

The Decision-Making Framework

When facing ethical questions about AI use in sales, use this framework:

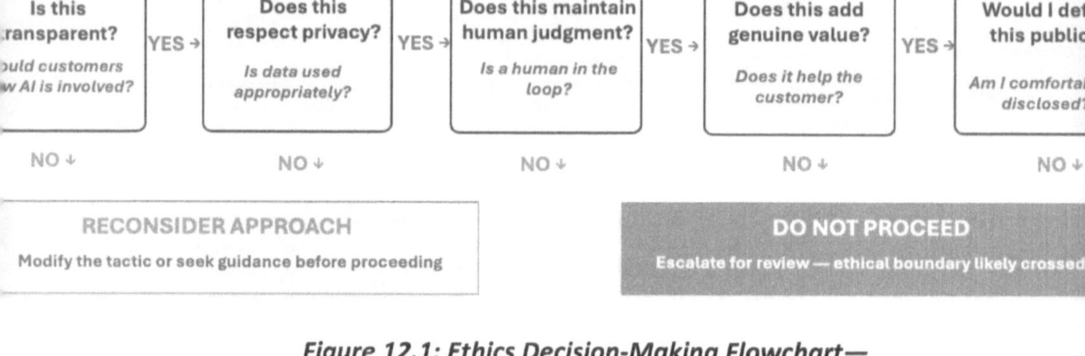

Figure 12.1: Ethics Decision-Making Flowchart—
Five Questions to Evaluate Any AI Sales Tactic

Question 1: Would I be comfortable if this were public?

If you'd be embarrassed or defensive if your practices were featured in a news article, that's a strong signal you're crossing ethical lines.

If you wouldn't want to explain your AI tactics to prospects, reconsider them.

Question 2: Does this respect the prospect's autonomy?

Are you helping them make informed decisions or undermining their ability to decide in their own interest?

Respecting autonomy: Providing information, addressing concerns, helping them evaluate options **Undermining autonomy:** Exploiting biases, creating pressure, deliberately obscuring information

Question 3: Would this build trust if they knew about it?

If prospects understood exactly how you're using AI, would they trust you more or less?

Trust-building: Using AI to better understand their needs, provide more relevant information, save them time **Trust-destroying:** Using AI to manipulate, survey, or exploit them

Question 4: Is this how I'd want to be sold to?

The golden rule applies to sales. If you'd feel uncomfortable being on the receiving end of your tactics, don't use them.

Question 5: Does this serve their interests or just mine?

Looking closer, the most important question. Ethical sales align your interests with the customer's. You succeed by helping them succeed. If you're prioritising your commission over their outcome, you're in dangerous territory.

Building an Ethical Sales Culture

Individual ethics matter, but organisational culture matters more:

Leadership Sets the Tone

If leadership celebrates wins at any cost, pressures reps to hit numbers regardless of how, or looks the other way at ethical corners being cut, the culture will be toxic.

Ethical leadership:

- Celebrates wins that came from genuine value creation
- Calls out unethical behaviour even when it was "successful"
- Models ethical decision-making publicly
- Creates safe channels for reporting ethical concerns

Incentives Drive Behaviour

If you only measure and reward revenue, you'll get revenue-at-any-cost behaviour.

Balanced incentives:

- Customer satisfaction and retention metrics alongside revenue
- Ethics and compliance as part of performance reviews
- Long-term value creation over short-term extraction
- Consequences for unethical behaviour that outweigh short-term gains

Training and Support

Reps need guidance on ethical AI use:

- Clear policies on what's acceptable and what's not
- Examples of ethical dilemmas and how to resolve them
- Regular training on evolving ethical issues

- Easy access to compliance and ethics support when questions arise

Technology Governance

Implement guardrails in your AI systems:

- Review AI-generated content for ethical issues before broad deployment
- Audit AI systems to ensure they're not discriminating or manipulating
- Establish approval processes for new AI tools and techniques
- Regular ethics reviews of how AI is being used in practice

The Long-Term Advantage of Ethics

Let me be clear about something: ethical selling isn't just about doing the right thing (though that matters). It's also smart business strategy.

Ethical sellers outperform over time because:

Trust compounds: Every ethical interaction builds trust. Over years, this creates reputation and referrals that are impossible to compete with.

Relationships last: Customers acquired ethically stay longer, expand more, and advocate more than those acquired through manipulation.

Talent stays: Good salespeople want to work for ethical organisations. Unethical cultures suffer high turnover.

Regulation favours you: As privacy and AI regulations tighten, ethical practices put you ahead of regulation rather than scrambling to comply.

Brand strengthens: Reputation for integrity is a competitive advantage that can't be copied quickly.

The salespeople who succeed long-term aren't those who exploit every possible advantage. They're those who build trust systematically by using powerful technology responsibly.

In the final chapter, we'll look ahead to where AI-powered sales are heading. The emerging technologies, evolving capabilities, and future trends that will shape how elite sellers operate in the years ahead.

CHAPTER OUTCOME

☐ A clear ethical stance on AI use in selling (what you will/won't do)

☐ A privacy and data-use rule-set (sources, consent, retention, access)

☐ An authenticity rule: what must always be human-owned (judgment, relationship, commitments)

☐ A transparency test you can apply to any AI-driven action ("would I be comfortable if public?")

☐ A list of red flags (manipulation, deception, unverifiable claims, misrepresentation)

☐ A content integrity process (fact-check, tone check, bias check)

☐ A risk plan for breaches/errors (how you respond and remediate)

CHAPTER 13

The Future of AI-Powered Sales

By the end of this chapter, you will be able to:

- Describe what's changing near-, medium-, and long-term (autonomous assistance → cognitive assistants → collaborative intelligence).
- Identify practical capabilities that are emerging (autonomous prospecting, real-time intelligence, predictive orchestration, hyper-personalisation).
- Anchor on what won't change (fundamentals, human connection, trust, commercial clarity).
- Build a personal readiness plan (skills, adaptability, staying current) without chasing shiny tools.
- Leave with a clear perspective on opportunity and how to stay ahead without losing the human edge.

Preparing for what's next while staying grounded in what works today

In 2010, if you'd told me that within 15 years, AI would transcribe sales calls in real-time, predict which deals would close with 85% accuracy, and generate personalised proposals for dozens of prospects simultaneously, I would have dismissed it as science fiction. Yet here we are. Those capabilities exist today and are becoming standard practice for leading sales organisations.

This raises an important question: what will the next 5-10 years bring? What capabilities that seem like science fiction today will be commonplace tomorrow? And more practically, how should you prepare for that future while still operating effectively in the present?

Consider that this final chapter looks ahead. Not to make definitive predictions (which are usually wrong), but to identify emerging trends, explore implications, and provide guidance on staying ahead of the curve without getting distracted by every shiny new technology that emerges.

The future of AI-powered sales isn't about replacing human sellers. It's about human sellers becoming exponentially more effective by combining human judgment with increasingly sophisticated AI capabilities.

The Current State: Where We Are Today

Before projecting forward, let's establish the baseline of what's possible right now in 2025:

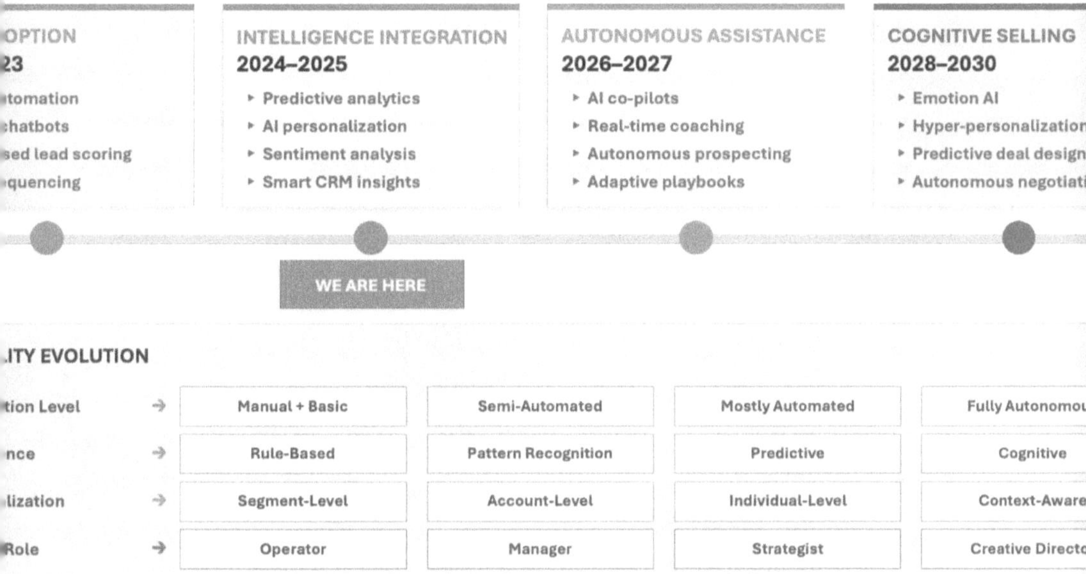

Figure 13.1: The Evolution of AI in Sales - Four Phases from 2020 to 2030 – From Basic Automation to Cognitive Partnership

AI can already:

- Transcribe and analyse sales conversations with high accuracy
- Score leads and predict deal outcomes based on behavioural patterns
- Generate personalised email content and proposals at scale
- Identify optimal send times and messaging for outbound sequences
- Map stakeholder networks and suggest relationship pathways
- Monitor for buying signals across digital channels
- Provide real-time coaching during sales calls
- Automate CRM data entry and activity logging
- Forecast revenue more accurately than human judgment alone
- Optimise pricing based on similar deal patterns

What AI still struggles with:

- Genuine creative problem-solving and strategy
- Reading subtle human emotions and interpersonal dynamics
- Building authentic long-term relationships
- Working through ambiguous situations with incomplete information
- Making judgment calls that require ethics and values
- Adapting to completely novel situations with no historical data
- Understanding context that requires lived human experience

In practice, the pattern is clear: AI excels at pattern recognition, data processing, optimisation, and automation. Humans excel at creativity, empathy, judgment, relationship-building, and steering through ambiguity.

Near-Term Evolution (2025-2027): Autonomous Assistance

The next 2-3 years will see AI move from tool to assistant. Not just providing information and suggestions but taking actions on your behalf with your oversight.

Autonomous Prospecting

Today: AI suggests prospects to target. You review the list and initiate outreach.

Near future: AI continuously monitors your ICP criteria and buying signals, automatically adds high-quality prospects to sequences, personalises initial outreach, and only escalates to you when prospects engage.

You'll shift from "working lists" to "reviewing AI-initiated conversations." Instead of spending time finding and contacting prospects, you'll spend time on the prospects AI has pre-qualified and engaged for you.

Implications: Reps who embrace autonomous prospecting will work 3-5x more opportunities with the same effort. Those who resist will struggle to keep up.

Real-Time Conversation Intelligence

Today: AI analyses calls after they happen and provides insights later.

Near future: AI provides real-time guidance during calls—surfacing relevant content, suggesting questions, flagging risks, recommending next steps—all while the conversation is happening.

Think of it like GPS for sales conversations: real-time navigation that keeps you on track toward your destination while adapting to unexpected turns.

Implications: Less experienced reps will perform like veterans because they'll have AI coaching them through every conversation. Training new reps becomes dramatically faster.

Predictive Deal Orchestration

Today: AI predicts deal outcomes and flags risks. You decide how to respond.

Near future: AI not only predicts what will happen but recommends specific interventions and orchestrates multi-step actions to improve outcomes.

"This deal has 38% close probability, down from 52% last week due to declining stakeholder engagement. I've identified 3 actions that historically improve deals in similar situations: 1) Schedule executive briefing with CFO within 7 days, 2) Share specific case study from financial services vertical, 3) Request technical validation meeting with CTO. I've drafted emails for each. Approve to send?"

Implications: Deals become actively managed by AI systems that never forget, never miss signals, and constantly optimise based on real-time data.

Hyper-Personalisation at Scale

Today: AI helps generate personalised content. You review and refine.

Near future: AI creates genuinely individualized content for each stakeholder that reflects not just their company and role, but their communication style, preferences, concerns, and behavioural patterns.

Each prospect gets an experience that feels custom-designed for them—because it is—without you spending hours per person creating it.

Implications: The baseline for "good" personalisation will rise dramatically. Generic outreach will become even less effective as prospects get used to hyper-relevant, hyper-personalised communication.

Medium-Term Evolution (2027-2030): Cognitive Sales Assistants

Looking 3-5 years ahead, AI will evolve from assistant to collaborative partner. Systems that genuinely understand context, anticipate needs, and engage in sophisticated reasoning.

Contextual Understanding and Reasoning

Future AI won't just process data. It will understand business context, strategic implications, and subtle situations:

"Your prospect just announced a CFO transition. Based on patterns from 200+ similar situations, new CFOs typically pause major purchase decisions for 60-90 days while they assess priorities. However, deals that were already in advanced stages often accelerate because new CFOs want quick wins. Your deal is in technical validation. Advanced enough to potentially accelerate. Recommend: 1) Acknowledge the transition in your next communication, 2) Position your solution as enabling quick, measurable win the new CFO can point to, 3) Offer accelerated timeline with additional support to make their first 90 days successful."

This isn't just pattern matching. It's contextual reasoning that considers multiple factors simultaneously.

Emotional Intelligence and Sentiment Analysis

AI will develop more sophisticated understanding of human emotions and interpersonal dynamics:

During calls: "Your champion's tone shifted when discussing budget. Vocal stress and hesitation suggest uncertainty despite positive words.

Recommend directly addressing budget confirmation before next stage."

In messaging: "Prospect's response time increased from 2 hours to 2 days, and recent emails are shorter and less enthusiastic. This pattern predicts declining interest. Recommend re-engagement strategy or re-qualification."

In stakeholder analysis: "While CTO is formally positive, meeting participation patterns and question types suggest underlying scepticism. Recommend one-on-one technical session to address unspoken concerns."

This moves beyond surface-level analysis to understanding what's really happening emotionally and politically.

Strategic Scenario Planning

AI will simulate complex scenarios and recommend strategies:

"If we pursue aggressive pricing to beat Competitor X, we have 68% probability of winning but at 30% lower margin. If we hold pricing and emphasise superior implementation support, we have 54% probability of winning at full margin. If competitor wins and implementation struggles (42% probability based on their track record), we have opportunity to displace them in 12-18 months. Recommended strategy: Hold pricing, invest in relationship for potential displacement opportunity, but have aggressive pricing fallback if deal is must-win for strategic reasons."

Recognise that this is sophisticated strategic thinking. Weighing multiple factors, considering probabilities, planning for contingencies.

Autonomous Negotiation Support

AI will actively participate in negotiation strategy:

"Prospect requested 25% discount. Based on similar deals, this is opening position. Typical final discount in this segment is 12-15%. Recommend: counter with 8% discount plus enhanced support package valued at 5% (actual cost to us: 2%). This positions us as responsive while preserving margin. If prospect pushes back, these 3 contract terms are typically negotiable in exchange for holding price. I've prepared alternative proposal structures for each scenario."

AI becomes a real-time negotiation advisor with deep pattern knowledge.

Long-Term Evolution (2030+): Collaborative Intelligence

Looking beyond 5 years becomes increasingly speculative, but we can identify likely trajectories:

AI Sales Partners, Not Just Tools

The distinction between "using AI" and "working with AI" will blur. You won't just use tools. You'll collaborate with AI systems that have deep understanding of your customers, your products, your strategies, and your personal working style.

"Good morning. Based on overnight developments, three priorities for today: Acme Corp's CEO posted about Q2 challenges. Perfect timing to re-engage with your value proposition on efficiency. TechStart Inc's technical validation is complete—recommend scheduling pricing conversation this week while momentum is high. Global Systems deal has stalled—I've analysed the situation and believe the issue is stakeholder alignment, not solution fit. Let's discuss strategy."

This is AI as colleague, not software.

Predictive Customer Understanding

AI will predict customer needs before customers themselves articulate them:

"Based on TechCorp's growth trajectory, technology investments, and hiring patterns, they'll need sales automation capabilities in Q3 2026 with 73% confidence. They're not actively looking yet, but early engagement now positions us as trusted advisor rather than just vendor responding to RFP. Here's a thought leadership piece addressing challenges they'll face at their projected scale."

This shifts from reactive (responding to expressed needs) to proactive (anticipating future needs).

Multi-Modal AI Integration

AI will naturally integrate across all communication modes. Voice, video, text, in-person:

During in-person meetings, AR glasses or other interfaces provide real-time guidance visible only to you. Video calls include AI analysing body language and facial expressions. Voice calls transcribe and analyse in real-time. Text communications generate instantly with full context.

Notably, the mode of communication becomes irrelevant. AI augments all of them.

Collective Intelligence

Individual AI assistants will share learnings across organisations (with appropriate privacy protections), creating collective intelligence that improves for everyone:

"In the past month, 47 companies in financial services changed their evaluation criteria to prioritise integration speed over feature breadth. This pattern suggests industry-wide shift. Recommend adjusting your financial services messaging to emphasise integration efficiency."

You benefit from patterns across thousands of sellers, not just your own experience.

What Won't Change

Amidst all this evolution, some fundamentals remain constant:

People buy from people they trust. No matter how sophisticated AI becomes, trust still requires human connection. AI can enable that connection to happen faster and at scale, but it can't replace it.

Genuine value creation matters. AI can help you communicate value more effectively, but if your solution doesn't solve customer problems, no amount of AI will create lasting success.

Ethics and judgment remain human. AI can optimise tactics, but strategic decisions about how to build relationships, when to walk away from bad-fit customers, and how to balance short-term revenue with long-term reputation require human judgment.

Creativity and innovation stay human-driven. AI can optimise existing approaches, but breakthrough strategies, novel solutions to unique problems, and creative deal structures still come from human ingenuity.

Emotional intelligence is uniquely human. AI can analyse emotions, but genuine empathy, reading a room, adapting to interpersonal dynamics. These remain human capabilities.

The future isn't about AI replacing sellers. It's about AI handling everything that can be automated, predicted, or optimised. Freeing humans to focus on what they do best: building relationships, applying judgment, and creating value.

Preparing for the Future

How do you prepare for this evolving ecosystem without getting paralysed by uncertainty?

Build Strong Fundamentals

The basics matter more, not less. As AI handles mechanical tasks, the differentiator becomes how well you do what AI can't: build relationships, exercise judgment, create strategy.

Invest in:

- Communication skills and emotional intelligence
- Strategic thinking and business acumen
- Industry expertise and domain knowledge
- Relationship building and networking
- Ethical decision-making and judgment

These capabilities compound over time and become more valuable as AI handles everything else.

Stay Technologically Current

You don't need to be on the bleeding edge, but you can't afford to fall behind:

Practical approach:

- Follow major AI sales vendors and their roadmaps
- Experiment with new capabilities as they enter mainstream
- Attend industry conferences and webinars on AI in sales
- Join communities where leading sellers share what's working
- Budget time and money for continuous learning

Aim to be in the "fast follower" category. Not first to adopt everything, but never more than 6-12 months behind leading edge.

Develop AI Collaboration Skills

Learn to work effectively with AI:

- **Prompt engineering:** How to give AI context and direction to get useful outputs

- **Critical evaluation:** How to assess AI suggestions and know when to override them
- **Feedback loops:** How to help AI systems improve by providing good training data
- **Integration thinking:** How to combine AI capabilities across tools effectively

For your team, these meta-skills become increasingly valuable as AI capabilities expand.

Build Adaptability

At its simplest, the specific tools and techniques will change. The ability to adapt won't:

Cultivate:

- Comfort with experimentation and learning from failure
- Openness to changing approaches when evidence suggests better ways
- Resilience through technological disruption
- Willingness to unlearn old habits that AI makes obsolete

The most successful sellers in 2030 won't be those who mastered 2025 techniques. They'll be those who continuously evolved as capabilities changed.

Maintain Human Connection

As AI handles more tactical work, double down on human elements:

Prioritise:

- In-person meetings and relationship building
- Executive access and C-level relationships
- Customer advisory boards and strategic partnerships
- Industry presence and thought leadership
- Authentic networking and peer community

These high-touch, high-trust activities become more valuable as routine interactions become automated.

The Opportunity Ahead

Here's my fundamental belief about the future of AI-powered sales: the opportunity is larger than the threat.

Yes, some things will become obsolete:

- Purely transactional sales roles where AI can fully automate
- Skills that AI replicates better than humans (research, data entry, basic personalisation)
- Organisations that resist AI and fall behind competitors

But more things will become possible:

- Individual sellers managing 3-5x more opportunities with same effort
- Smaller companies competing effectively against enterprises through AI use
- Faster sales cycles and higher win rates through better intelligence
- Deeper customer relationships through better understanding
- More fulfilling sales careers focused on strategy and relationships rather than administration

The salespeople who thrive won't be those with the highest natural talent or most aggressive tactics. They'll be those who best combine human judgment with AI capabilities. Creating a partnership between human and machine intelligence that's more effective than either alone.

A Final Thought

Throughout this book, we've explored the QUANTUM Framework for AI-powered sales: how to Qualify prospects intelligently, understand customers deeply, anticipate needs before they're voiced, guide organisational complexity, Tailor approaches at scale, unify systems into coherent ecosystems, and Measure what predicts success.

We've discussed tools and tactics, strategies and ethics. But ultimately, success in AI-powered sales comes down to something simpler:

Use technology to become more human, not less.

Let AI handle the mechanical, the predictable, the repetitive. Free yourself to focus on creativity, empathy, judgment, and relationship-building. The things that make you uniquely human and uniquely valuable.

Consider how the sellers who win in the AI age won't be those who hide from technology, hoping it goes away. They won't be those who blindly adopt every new tool without strategy. They'll be those who thoughtfully integrate AI capabilities to amplify their human strengths.

They'll work smarter, not just harder. They'll understand customers more deeply. They'll close deals faster. They'll build more valuable relationships. And they'll do all of this while maintaining their authenticity, their ethics, and their humanity.

That's the promise of AI-powered sales. Not replacing salespeople but releasing them.

The future is coming whether you're ready or not. The question is: will you shape it or be shaped by it?

The tools are available. The framework is proven. The opportunity is massive.

Your Next Steps

Now that you understand the QUANTUM Framework and the future trajectory of AI in sales, here's how to get started:

- **Assess your current state:** Review which QUANTUM stages you're executing well and which need attention. Most organisations excel at 2-3 stages but have gaps in others.
- **Start with one stage:** Don't try to implement everything at once. Pick your biggest constraint (usually Qualify, Understand, or Measure) and focus there first.
- **Invest in foundational tools:** Begin with conversation intelligence and CRM enhancement. These provide the data foundation everything else builds on.
- **Build the feedback loop:** Ensure you're capturing data, analysing patterns, and applying learnings. This continuous improvement is what separates good from great.
- **Stay ethically grounded:** As you adopt more sophisticated AI capabilities, regularly revisit the ethical framework in Chapter 12. Technology should amplify your humanity, not replace it.

Now it's time to get to work.